Cambridge Elements

Elements in Earth System Governance
edited by
Frank Biermann
Utrecht University
Aarti Gupta
Wageningen University
Michael Mason
London School of Economics and Political Science

INSTITUTIONALISING MULTISPECIES JUSTICE

Danielle Celermajer
University of Sydney
Anthony Burke
University of New South Wales
Stefanie R. Fishel
University of the Sunshine Coast
Erin Fitz-Henry
University of Melbourne
Nicole Rogers
Bond University
David Schlosberg
University of Sydney
Christine Winter
University of Otago

CAMBRIDGE
UNIVERSITY PRESS

CAMBRIDGE UNIVERSITY PRESS

Shaftesbury Road, Cambridge CB2 8EA, United Kingdom

One Liberty Plaza, 20th Floor, New York, NY 10006, USA

477 Williamstown Road, Port Melbourne, VIC 3207, Australia

314–321, 3rd Floor, Plot 3, Splendor Forum, Jasola District Centre, New Delhi – 110025, India

103 Penang Road, #05–06/07, Visioncrest Commercial, Singapore 238467

Cambridge University Press is part of Cambridge University Press & Assessment, a department of the University of Cambridge.

We share the University's mission to contribute to society through the pursuit of education, learning and research at the highest international levels of excellence.

www.cambridge.org
Information on this title: www.cambridge.org/9781009506236
DOI: 10.1017/9781009506243

© Danielle Celermajer, Anthony Burke, Stefanie R. Fishel, Erin Fitz-Henry, Nicole Rogers, David Schlosberg and Christine Winter 2024

This publication is in copyright. Subject to statutory exception and to the provisions of relevant collective licensing agreements, with the exception of the Creative Commons version the link for which is provided below, no reproduction of any part may take place without the written permission of Cambridge University Press & Assessment.

An online version of this work is published at doi.org/10.1017/9781009506243 under a Creative Commons Open Access license CC-BY-NC 4.0 which permits re-use, distribution and reproduction in any medium for non-commercial purposes providing appropriate credit to the original work is given and any changes made are indicated. To view a copy of this license visit https://creativecommons.org/licenses/by-nc/4.0

When citing this work, please include a reference to the DOI 10.1017/9781009506243

First published 2024

A catalogue record for this publication is available from the British Library

ISBN 978-1-009-50623-6 Hardback
ISBN 978-1-009-50625-0 Paperback
ISSN 2631-7818 (online)
ISSN 2631-780X (print)

Cambridge University Press & Assessment has no responsibility for the persistence or accuracy of URLs for external or third-party internet websites referred to in this publication and does not guarantee that any content on such websites is, or will remain, accurate or appropriate.

Institutionalising Multispecies Justice

Elements in Earth System Governance

DOI: 10.1017/9781009506243
First published online: November 2024

Danielle Celermajer
University of Sydney

Anthony Burke
University of New South Wales

Stefanie R. Fishel
University of the Sunshine Coast

Erin Fitz-Henry
University of Melbourne

Nicole Rogers
Bond University

David Schlosberg
University of Sydney

Christine Winter
University of Otago

Author for correspondence: Danielle Celermajer, Danielle.celermajer@sydney.edu.au

Abstract: Multispecies Justice (MSJ) is a theory and practice seeking to correct the defects that make dominant theories of justice incapable of responding to current and emerging planetary disruptions and extinctions. Multispecies Justice starts with the assumption that justice is not limited to humans but includes all Earth others, and the relationships that enable their functioning and flourishing. This Element describes and imagines a set of institutions, across all scales and in different spheres, that respect, revere, and care for the relationships that make life on Earth possible and allow all natural entities, humans included, to flourish. It draws attention to the prefigurative work happening within societies otherwise dominated by institutions characterised by Multispecies Injustice, demonstrating historical and ongoing practices of MSJ in different contexts. It then sketches speculative possibilities that expand on existing institutional reforms and are more fundamentally transformational. This title is also available as Open Access on Cambridge Core.

Keywords: multispecies justice, environmental justice, climate justice, environmental politics, environmental governance

© Danielle Celermajer, Anthony Burke, Stefanie R. Fishel, Erin Fitz-Henry, Nicole Rogers, David Schlosberg and Christine Winter 2024

ISBNs: 9781009506236 (HB), 9781009506250 (PB), 9781009506243 (OC)
ISSNs: 2631-7818 (online), 2631-780X (print)

Contents

1 What Is Multispecies Justice? 1

2 Multispecies Justice from Theory to Practice 14

3 Local Examples of Multispecies Justice 25

4 Multispecies Justice and Law 37

5 Planetary Institutions for Multispecies Justice 51

6 Conclusion 66

References 70

1 What Is Multispecies Justice?

1.1 A World of Multispecies Injustice

In 2018, volunteers from International Animal Rescue released a video, taken five years earlier, of a single orangutan fighting a bulldozer as it destroyed the Sungai Putri Forest in Borneo, Indonesia, which was his and his community's home. This orangutan was rescued, but the decimation of most orangutans and the destruction of the rainforests across Southeast Asia that are the lifeworld for myriad others, including Indigenous Peoples, continues at an accelerating pace. Ecocide such as this is not simply a crime committed by malevolent individuals; it is the inevitable result of capitalist markets driving the annihilation of habitat and life as they seek out opportunities to create profitable monocultural environments better suited to the production of commodities that can flow through global supply chains. It is this same dynamic that is turning the Amazon, home to the greatest range of biodiversity and the most important carbon sink on the planet, into a net carbon emitter, as ancient forests, and the biocultural worlds they hold, make way for cattle ranches and soy plantations that will provide 'feed' for pigs in China and more cattle in the United States. The same pathology feeds forest fires across hemispheres year-in-year-out as mean temperatures rise, desiccating forests and soils. In Australia alone during the summer of 2019–2020, over twenty-four million hectares of bush were incinerated (Binskin, Bennett, & Macintosh 2020) and over three billion animals[1] perished (van Eeden et al. 2020; Christoff 2023). Meanwhile, during the Covid-19 pandemic, the mainly Latino, Black, immigrant and impoverished workers in overcrowded and unhygienic slaughter-houses across the United States fell ill and died at far above average rates. Justified by the 'constrained circumstances' for rapidly killing the pigs, chickens, and cows that could not be 'processed' through the meat production system, regulators approved extreme methods like shutting off ventilator systems and adding CO2.

In each of these cases, animals, environments, and the most marginalised humans are caught in networks of violence, destruction, and injustice. And this is no coincidence. Together, but also differently, their lives, worlds, and relationships are made expendable by the logics and institutions of contemporary global capitalism and persistent racism and colonialism. Scaled to the level of the planet, those logics and institutions are driving the simultaneous and linked crises of biodiversity, mass extinction, deforestation, global climate change, ocean acidification, destruction of Indigenous Peoples and cultures,

---

[1] Although humans are also animals, throughout this text, the term animals will refer to animals other than humans.

and pervasive social injustice, including anti-Blackness and the ongoing decimation of the Global South. Addressing them across all scales from the local to the planetary requires understanding them as multispecies *injustices*, and radically transforming institutions according to the logics and principles of multispecies justice (MSJ).[2]

In recent years, theorists, including the authors of this work, have laid out some of the core theoretical commitments of MSJ, summarised in this section. Our goal in this Element is to build on this work to *describe* and *imagine* a set of institutions, across all scales and in different spheres, that respect, revere, and care for the relationships that make life on Earth possible and that allow all natural entities, humans included, to flourish. By *describe* we mean to draw attention to the prefigurative work that is already happening within societies otherwise dominated by institutions characterised by multispecies *in*justice, at the same time as demonstrating the historical and ongoing practices of MSJ in a range of different sociocultural contexts and worlds. By *imagine* we intend to sketch out further speculative possibilities, ones that expand on existing institutional reforms and are more fundamentally transformational. In this regard, some of the institutional suggestions we make will be of a reformist and shorter term character, working within the existing institutional architecture, and to this extent, not seeking to directly deconstruct Westphalian categories such as the territorial *state*. At the same time, recognising that this architecture is not only historically contingent but in many ways also bound up with the understandings of justice we are criticising, we suggest more transformative changes, which use MSJ to change the architecture itself. Throughout the following sections, we offer examples of both reformist and transformational institutional change at multiple levels, in both the Global North and the Global South, from local democratic practices to new models of planetary institutions that are not organised around nation states.

Because they concern *justice*, the ethical commitments of MSJ constitute obligatory requirements for the basic institutions of society, not defeasible options or discountable side-constraints. This does not mean that MSJ proscribes an absolute set of abstract and universal rules that can or should be applied across all contexts. Rather, MSJ establishes a set of principles, logics, and considerations that must be taken seriously and inform the design and operation of governance and institutions within the particular contexts and sets of relationships where they are being negotiated. Those principles, logics, and considerations will not be the only ones that need to be part of just institutional design, and they will need to be brought into conversation with others, including those that are apparent only

---

[2] Throughout the Element, we use the acronym MSJ.

within the place-based and temporal situation where the decisions are being made. Our aim is not to set out rules to follow but to offer some ideas and provocations for a broader conversation about the myriad institutional forms at a range of scales and across various spheres of justice (judiciaries, municipalities, states, international organisations). We aim to put forward reforms and transformations that can support practices of justice that are appropriate to this era and are capable of overcoming the anthropocentrism, parochialism, and presentism that are impeding justice for oppressed and marginalised communities, for future generations, and for the future of the planet.

1.2 What Is Multispecies Justice?

1.2.1 Core Commitments

Multispecies justice must be understood as a theory *and* practice of justice emerging in the context of the multiple interlinked crises (that we will refer to as the polycrisis) of the twenty-first century.[3] It provides a critical analysis of the most dire injustices of our times, and aims to correct the defects that make dominant theories of justice incapable of responding to current and emerging planetary disruptions and extinctions. Indeed, MSJ suggests that dominant theories of justice have *contributed to* those conditions and offers an alternative ontological and ecological approach.

It does so in several respects. Here, we highlight three. First, MSJ embraces an understanding where what is primary is not individuals, but *relationships*. In technical terms, it adopts a relational ontology. This approach includes rejecting the idea that humans or human individuals exist as radically distinct and at the top of a hierarchy of being. Second, MSJ adopts an ethics and politics of difference. This means that it rejects anthropomorphism, or the idea that the capacities, affordances, and ways of being of humans (or humans as they have been imagined in any particular way) provide the 'ideal' or norm against which other beings ought to be evaluated. Third, MSJ tries to uncover the logics that undergird and connect different manifestations of injustice among humans and against beings other than humans. We explore each of these in more detail in what follows.

The first fundamental feature of MSJ is that it starts with an acknowledgement of the ontological and ethical primacy of the ecological relations within which all beings are enabled to, or inhibited from, functioning and flourishing (Kurki 2020; Celermajer 2021). This acknowledgement also means rejecting the picture of the human individual (who has been the classical subject of liberal

---

[3] Polycrisis can be defined as 'a series of interconnected and interacting threats – climate change and ecological disasters, rising economic inequality and political polarization, violent conflict and more' (Hoyer et al. 2023, cf also Lawrence, Janzwood & Homer-Dixon 2022).

justice) as radically separate, distinct from, and superior to the more-than-human world, which they are putatively entitled to exploit as means to the ends they establish for themselves. Multispecies justice does not simply recognise all Earth beings as subjects of justice; it embeds that recognition in a theory of being that places relationships first (a relational ontology) and enacts a non-hierarchical ethics.[4] In this regard, MSJ takes inspiration from Indigenous philosophies (Whyte 2013; Yunkaporta & Shillingsworth 2020), ecological justice theories (Schlosberg 2007), and various trends in posthumanism (Haraway 2016), as well as an appreciation of the symbiotic structure of all life or being (Margulis & Sagan 2013) that is now, belatedly, acknowledged across a range of natural and social scientific and humanities disciplines.

The primacy of this relational ontology and ethics has critical implications for how MSJ understands justice and for the types of institutional forms that would most fully realise it. First, justice is not, or certainly not *only*, concerned with the interests of individuals, or individual species, or categories of beings, as if they could be abstracted and then treated apart from the relationships that enable or impede their flourishing. This does not mean that individuals do not matter, but their existence is always (re)cast in a relational mode. This move may not seem very radical when applied to beings other than humans; indeed, within dominant western paradigms, Earth others[5] have often been characterised as part of undifferentiated resources or materials. In dominant theories of justice, the apparent 'failure' of Earth others to be adequately 'individual' or sentient has been used as a pretext to claim that they cannot count as subjects of justice. Multispecies justice insists that an adequate understanding of justice for all beings must acknowledge that they are all dependent on, and threatened or supported by, the relationships in which they are embedded (see, for example, Nedelsky 2011). At the same time, the relational ontology at work here is neither flat, homogenous, nor without differentiations. Rather, as posthumanists like Barad (2007) have insisted, differentiation is a function of relationships, and hence the quality of those relations and the way in which power flows through them, in the form of (for example) care, exploitation, or domination, shapes what is possible for differentiated individuals.

---

[4] The term anthropocentrism has several different meanings. Here, it is not what Baird Callicott (2013) calls 'tautological anthropocentrism', meaning experiencing from the perspective of a human, that we reject, but the placement of the human at the ethical centre and summit.

[5] Throughout this text, when we are referring to beings other than humans, and wishing to make that distinction, we will use the term Earth others. By Earth others we mean everything (all beings and relationships, excepting humans) that constitutes the totality of the Earth system: living and non-living elements that together make up this planet. Where we mean the totality, including humans, we use the term more-than-human. Where we are referring not to the totality but to all members of that totality, including humans, we use Earth beings.

Multispecies justice's embrace of a relational ontology also exposes the incoherence of the idea that justice for humans could be considered in isolation from or, even worse, on the basis of the exploitation of Earth others. Western disciplines across the humanities and social and natural sciences have, as noted, increasingly recognised the symbiotic or sympoetic nature of life, including human life, and today the truth of interdependence is flagrantly manifest in the devastation that ecological and climate crises, including pandemics, are bringing to human life (Celermajer & McKibbin 2023). As the absolute dependence on a functional ecological system for fundamental human needs – to eat, to drink, to breathe, to a decent standard of health, to stable and non-violent social relations – is now made clear on a daily basis, so too is the absurdity of theories of justice that assume that justice begins when humans depart from the 'state of nature'. Multispecies justice re-embeds justice in the relationship between nature and culture – a relationship which humans do not need to *leave* to experience or promote justice, but to which they need to attend.

Along with the rejection of anthropocentrism or human exceptionalism, MSJ rejects anthropomorphism, a worldview that takes 'the human', understood in a particular way, as the basic model of the subject of justice, and then ranks and evaluates all other beings according to how they compare. In this sense, MSJ is not an extensionist theory. Thus, MSJ does not adopt the logic of theories of justice that include some beings other than humans (principally certain other animals; see, for example, Singer 1975 and Regan 2004) by arguing that they should be accorded a certain ethical status because they possess those capacities deemed to qualify humans as subjects of justice, like a certain type or 'level' of sentience, reason, or language.

The rejection of anthropomorphism also entails the rejection of the superiority of some particular understandings of 'the human'. More positively, it entails embracing pluriversal[6] understandings of the human (and humans' relationship with Earth others). Insofar as MSJ also attends to the injustices involved in epistemic hierarchies, it must consciously avoid committing injustices by excluding particular ways of knowing or being and thereby pre-emptively precluding those understandings becoming part of the idea of MSJ. Multispecies justice has been inspired by the ethics and politics of difference, and the rejection of assimilative logics articulated across critical feminist, racial justice, Indigenous, crip, and queer theories (see Coulthard 2014; Joshi 2011; McRuer 2006; Medina 2012; Young 1990). There is of course a caveat to this open and pluriversal inclusivity: MSJ, a *critical* theory, openly and deliberately rejects politics of domination such as

---

[6] Pluriversalism eshews the hegemonic assumption that there is just one 'right' way of understanding the world and human positionality. The pluriversal vision is for radical openness to multiple possibilities for knowledge making and understanding what it is to be human and/or multispecies.

some manifestations of liberalism (such as capitalist neoliberalism and late liberalism), neo-fascisms, extractivism, and colonialism. These world views and politics are the subject that MSJ seeks to critique and away from which it seeks to turn. Put differently, they are one causal spur to the institutionalisation of MSJ, and their totalising and universalising impulses are those which MSJ both spurns and deconstructs.

While on first blush MSJ would seem to be primarily (if not exclusively) concerned with injustices committed against Earth others, it is in fact equally concerned with a variety of intra-human injustices, including racism, colonialism, and gender discrimination. This is because MSJ understands these different sites of injustice as logically and institutionally connected. Thus, MSJ's third core commitment is to draw connections between different kinds of injustice that are often seen as distinct, including instances of injustice against different groups of humans and Earth others. This theoretical commitment entails a practical one of developing strategies that address the pathological logics that undergird different but connected injustices. As ecofeminists, critical race theorists, and critical disability theorists have long articulated, the systematic exclusion or marginalisation of certain humans because of their racialised, gendered, or disabled identity is patterned by a set of constructed and hierarchically organised binaries rooted in the ideal of a certain type of human (Plumwood 2002; Taylor 2017; Wynter 2003). These hierarchical binaries legitimise violence against and exploitation of the 'less than perfect' human and more-than-human. For MSJ, violence against Earth others cannot be addressed without also confronting pervasive, entrenched and systematic anti-Blackness, ongoing coloniality, and gendered violence (among other injustices). By the same logic, these intra-human forms of violence cannot be challenged without appreciating how they are produced through understandings of 'the human' based on radical separation from, superiority to, and domination over what is not human, as well as through the view that this 'not-human' is an extractable, exploitable resource for those who count as fully human.

1.2.2 The Need to Clarify Both Root Terms: Multispecies and Justice

The basic terms that comprise MSJ – *multispecies* and *justice* – may appear self-explanatory. Each is, however, multifaceted and contested and therefore we, the authors, describe here how each is conceived, the stakes in conceptualising them in different ways and how we will be using them.

To begin with the 'species' in multispecies, we recognise an internal contradiction in our use of this Linnean-derived concept, developed to divide the world into component parts and *rank* them according to their attributes (see

Celermajer et al. 2020; Haraway 2008; Van Dooren, Kirksey, & Munster 2016). The problem with the term is further exacerbated by the dominance of the idea of a hierarchy of species, with humans at the top, an imaginary that has long legitimated and rationalised domination over and abuse of Earth others and those (humans) who are deemed not to meet the human 'ideal' (Pellow 2016; Kojola & Pellow 2020). Moreover, 'species' has generally referred only to living beings, whereas, as we will discuss shortly, our view is that the life/non-life boundary should not be imported into MSJ. As such, including the word 'species' may well seem the antithesis of the open ontology the theory aims for. It is a tension that scholars are attempting to navigate, while increasingly using the term in a range of critical and reconstructive ways (Youatt 2022; Alberro 2024).

Then there is the question of which beings count as the 'species' who qualify for MSJ. Some take ('higher order') sentience as the marker of inclusion (Nussbaum 2007), others include all animal beings (Meijer 2019; Crary & Gruen 2022), and still others take life as the criteria and thus include plants. Still others add ecosystem flourishing, all matter and elements to this list (Watene 2016; Winter 2022a; Winter & Schlosberg 2023). Our definition of multispecies includes all Earth others – not only non-human animals, plants, fungi, and protists but also the elemental (to adopt the categories assumed in contemporary western thought). The problem is that as we recorded those positions you will have detected the lingering but powerful hold of Linnean categorisation.[7] In taking a pluriversal and inclusive stance, MSJ eschews hierarchies and is cautious of how boundaries (such as the species boundaries between animals and plants, life and non-life) are drawn. Rather, it accepts that all existence and the existence of all on Earth are dependent on relationships between living and non-living, plant and animal, material and elemental being.

The variant of MSJ we describe here, what we might describe as a thoroughly inclusive theory of MSJ, has arisen from a continuum of thought. As with all intellectual exercises it is part of an intergenerational project that builds from its antecedents while responding to emergent challenges. While in Indigenous philosophies the various distinctions living/non-living, animal/plant/fungi, higher/lower, and so on are not hard placed markers of degrees of human responsibility (Whyte 2013; Watene 2016; Yunkaporta & Shillingsworth 2020; Winter 2022a, 2023), MSJ in its academic form is emerging from centuries of such distinctions. When Donna Haraway coined MSJ in *Staying*

---

[7] We should also note that while science divides organisms into species analytically, all organisms (including humans) coexist with many other species, both within their own bodies and in ecosystems (Fishel 2017). It is also clear to science that the categories are themselves simply a useful tool; they do not describe the living world as it actually is (Wilkins 2018).

with the Trouble (2016), it was done within that emerging intellectual milieu, when the boundaries between biological and more specifically animal 'species' were brought into question. Some theorists prefer to maintain this animal distinction, others include all living things and others all Earth beings. We recognise that the term carries the legacies of the distinctions entailed in the language of species, both among living beings and between living and non-living beings. What we grapple with here and invite others to grapple with is whether, given the traction the term MSJ has gained, it is worth jettisoning for another more accurate label. Here we have chosen to build on the momentum the term has garnered rather than attempting to coin a new term or appropriate an old one. While an alternative may be more accurate, it will take time to become known and accepted (if it ever is) and potentially impede the critical shift in thinking and action that the field, thus named, is nourishing. Here, we take MSJ as a nomenclature for an inclusive, pluriversal, and adaptable relational conception of justice – with an understanding of the important and valuable arguments around the term.

In suggesting that the 'species' of MSJ includes beings across this full range of categories, we are not arguing that they all need to be included in every instance. Indeed, finding the most just way forward in any concrete situation will involve difficult trade-offs between different justice claims within this enlarged community of subjects. Nevertheless, to be capacious and pluriversal, and to avoid committing its own epistemic injustices by excluding the worldviews of certain peoples, a theory of MSJ must not *foreclose* worldviews and ontologies that, in different ways, also reject anthropocentrism and extractive relations with Earth others. In this regard, questions of inclusion and exclusion are not simply ontological, that is, questions about the 'real' nature of types of beings, but also ethical and political, questions about what different classificatory systems enable or impede in terms of how humans relate to others.

There is too the problem of a collective noun for the (the adjective) *multispecies*. The English language provides few possibilities. There is 'the environment', a term too imbricated in dualism and with notions of wilderness, background and homogeneity (see Morales 2019). Moreover, environmental justice has come to refer to a very specific set of injustices faced mainly by people of colour, poor people, and Indigenous Peoples, legitimated, non-coincidentally, by the hierarchies of humanness that place the aforementioned groups closer to the animal. Non-human is frequently used, but problematically characterises Earth others through their lack and reinscribes the human-other dualism. There are attempts to find the most suitable term here, some that have been used for decades, like ecological (Baxter 2004); others that are newer in academic usage – planetary (Biermann & Kalfagianni 2020; Sultana 2023), multi-being (Reid 2023), other-than-human and Earth being (De la Cadena 2015).

Our preferred term for the collective noun for multispecies is more-than-human. More-than-human encompasses all Earth beings, humans included, and understands them as entangled. As such, it is consistent with the relational ontology we have laid out. Importantly, humans are included insofar as they are, like all others, understood as belonging to a larger set of relationships. Nevertheless, precisely because it is Earth others and particular humans who have systematically been excluded from dominant western theories and institutions of justice, it will often be necessary to explicitly spell out who is included within this collective term, and, when it comes to institutional reforms, to explicitly attend to the conditions of different beings' flourishing. We will also, throughout this Element, use a second term – Earth others – when we are referring to the full collective of the more-than-human, but excepting humans. We find it necessary to have two terms: more-than-human, which is fully inclusive, and Earth others, which excepts humans, as in some instances we need to point to the institutional and ethical distinction that carves humans out as a distinctive category.[8]

Whatever the hesitations, the idea of MSJ, coined by Haraway, currently resonates across disciplines, is employed in a substantial body of literature that draws attention to the damages wrought by human/non-human dualisms, and focuses on justice across categorical divisions (e.g. Chao, Bolender, & Kirksey 2022). The term has gathered its own force, including through our own work. At a moment of environmental instability and destruction, it seems inappropriate to divide attention and the momentum of good the term has generated in the search for a more perfect term. The point is to imagine an inclusive, relational ecological world as the shared community of justice.

Turning to the term justice, we take justice to be a necessary character of institutions. In this Element, while acknowledging the multiple existing approaches to justice and the creative work going on in other justice spheres, our objective is to bring to them a multispecies lens, demonstrating the potential fullness of earthly flourishing in its diversity and difference under a collective practice of MSJ. We are interested in the various dimensions across which justice has been conceptualised – distributional (who gets what), recognition (whose identity is recognised and valued), participation (who gets a say and how), capabilities (what is necessary for functioning), and also reparative (how

---

[8] An indication of the difficulty in finding the right terminology here is that the authors of this Element could not fully agree on which terms to use. Specifically, some prefer the term other-than-human to Earth others. In this regard, we recognise that others will prefer different terms. Earth others is where we settled.

to be fully responsible for past wrongs), transitional (how to justly transition societies from periods of systemic injustice), and epistemic (whose knowledge counts). Again, our purpose is to bring to the conceptualisation and practice of justice across these dimensions a multispecies lens that can move them beyond the (individual) human being as the subject of justice and justice's addressee as the state, whose duty is to support (or at least not to frustrate) legitimate human life projects and provide the conditions that allow for human flourishing. Our understanding of justice in this regard is developing in the context of the polycrisis, which demands more complex, multi-scalar, multidimensional, and connected intrastate frameworks and accountabilities, which can marshal local, national, and the international community and institutions to collective responses for planetary stabilisation at the least, thriving at best.

For clarity, we regard 'justice' as a human responsibility. The obligations and duties of justice are a guide to human politics and institutional behaviour. Multispecies justice clarifies for politics the breadth of the responsibilities owed for all earthly flourishing. This does not make MSJ anthropocentric, because while they are its targets, human beings are no longer the sole regard of justice.

1.3 The Roots and Relations of MSJ

While it is tempting to describe MSJ as 'new', it is at once millennia old and contemporary in Earth-centric communities. Throughout time and across the globe, people and peoples have philosophised about and developed practices and protocols for the obligations and duties the idea demands. That knowledge, the epistemic frameworks, and, importantly, the ontological underpinnings have been largely quashed in European/Anglo/American intellectual communities and governmental practices. They remain central to many, if not all, Indigenous peoples' philosophies, epistemologies, ontologies, and, critically, governing practices (Alfred 1999, 2005; Winter 2022a). Each community has different sets of underlying explanations for, and practices and protocols governing, their understanding of living in harmony with and fostering the flourishing of the other beings with whom they share spaces, derived from their specific epistemologies, ontologies, social and political organisations, and the environments and ecosystems towards which they bear responsibility. An academic theory of MSJ can acknowledge and learn from such practices, holding its own boundaries sufficiently wide to ensure a place for Indigenous ways of being and justice *to* Indigenous Peoples and people, and at the same time avoid practices of extraction and exploitation. The practices of MSJ in Indigenous societies should not be wrenched from the entangled and emplaced relationships, epistemologies and ontologies from whence they grew and where they are maintained.

At the same time, as an emergent theory and discourse within the western academy, MSJ draws critical insights from and ought to be seen as an outgrowth of several other theories of justice. Specifically, MSJ has roots in several decades of ecological justice theory and activism as well as animal justice and critical animal studies (see Celermajer et al. 2021). Over the past decades, scholars and activists committed to environmental and ecological justice have included more-than-human flourishing in their domains of concern (see Gleeson & Low 1998; Plumwood 1999; Schlosberg 2007, 2013, 2014; Whyte 2016; Pellow 2017; Kojola & Pellow 2020), and people of colour and Indigenous Peoples have stressed the connection between environment, culture, and identity and commitments to a thriving whole (Whyte & Cuomo 2017; Whyte 2019). Within these conceptions of justice lie concepts of sacredness, ecological unity, and an understanding of, and commitment to care for entanglements between all Earth beings. Ecological justice addresses concerns of justice for ecological systems by noting, for instance, that the integrity of the ecological system is a matter of justice – for people and the system (Schlosberg 2013, 2014).

Multispecies justice also draws on critical insights of a range of justice theories that attend to the ways even the more progressive approaches to justice – those developed from liberalism's devotion to equality and fairness – continue to dominate and oppress some humans and Earth others. These include ecofeminism, critical race and disability theories, and various posthumanisms. Carole Pateman (1988), Charles Mills (1997), and Stacy Clifford Simplican (2015) have exposed the sexual, racial, and 'capacity' domination within the social contract – the very idea that underpins liberal state legitimacy. Specifically, the social contract and dominant justice theories have been designed to work for white, heterosexual, cisgendered, able-bodied men (Opperman 2022). They do not sufficiently support women, racialised, colonised, queer and disabled-bodied human subjects, nor the totality of the other-than-human realm. They largely ignore the different and differentiated needs of those who fail the test of the ideal human individual and erase the often-exploitative conditions of possibility for the individual to flourish: the social contract is then a racial–sexual–speciesist–ableist–colonial contract (Winter 2022b). The consequent exclusion of each of these categories of being from the spheres of justice is referential: the racialised or gendered person is animalised, or an Indigenous philosophy expressed within deep ecological engagement and learning is trivialised as fantasy, myth, story, and so forth. The harms done to each subset of being are thus both networked and exponentially exacerbated. Simultaneously, insofar as MSJ is part of a response to the global crises born of social and economic institutional forms and imaginaries that prescribe scarcity and competition as the endpoint of human political arrangements, MSJ is inspired by and seeks to work in alliance with anti-colonial justice, anti-capitalist justice (especially in its eco-socialist forms), anti-racism, anti-neoliberalism and other justice theories, and movements that imagine and seek to construct alternatives to social and political orders that are based on hierarchical ontologies and that perpetuate domination (Sultana 2021a, 2021b; Tschakert et al. 2021; Alberro 2024).

Critiques of the ways in which liberal justice privileges the present have also influenced the development of MSJ. In most formulations, the subject of justice is identifiable and alive. The long-term temporal unravelling of climate change and the polycrisis has brought the idea of intergenerational justice into sharper focus; however, although many have tried to include future generations within the existing justice framework, it largely remains mired in problems of individual identity (Caney 2010; Winter 2022a). Some do allow for expansions to responsibility for overlapping generations within a polity (de-Shalit 1995; Page 2007); however, the consequences of current fossil fuel burning, terraforming, and resource extraction will be felt thousands of years into the future – that is, across many more-than-human future generations. Furthermore, although a feature of Indigenous worldviews, few political theorists have attempted to include current generations' obligations or duties to ancestors (exceptions include Kumar 2003 and O'Neill 1993). Multispecies justice attempts to navigate beyond the individual and beyond any asymmetric favouring of the present for the sake of future earthly flourishing.

1.4 Why Multispecies Justice Now?

The long-term abuse of some humans and Earth others, or more broadly what we call the more-than-human, provoked the idea of MSJ, and its theorisation has grown in depth and width from its earlier articulations. So uninterrupted is the flow of unnatural disasters that communities (human and ecological) have no time to recover from one event before the next hits – a condition of turbulence (Dauvergne & Shipton 2023). Multispecies justice addresses the woven threads of these disasters by identifying their root problems as a set of interlinked cultural, economic, political, and legal institutions that combine to divorce 'the human' from all Earth others and erase the richness of more-than-human lives and the complexities of relationships embedded in both the living and non-living realms, and to cast Earth others as economic resources. It is this institutional web that supports the unbridled exploitation of the more-than-human for capital gain. For the most part, even regulations to 'protect' the environment are focused on the value of the more-than-human *to humans*, which in turn facilitates logics of financial calculation to determine whether to exploit or protect.

By glorifying the human subject and eliminating responsibilities and duties to those deemed 'objects' or 'resources', institutions and justice theory have actively undermined the conditions of well-being they proport to support. The schema of human-as-dominator has had and continues to have a hold over a range of regulatory regimes. It maintains conditions of ongoing colonialism, neocolonialism, and racism, which hold at their core a conviction that some people, places, species, waters, and atmospheres are disposable (Sassen 2014). As the devastation of these logics becomes more visible (and not only to the colonised and racially marginalised, to whom they have long been evident), as changes in the climate and Earth systems routinely disrupt lives and communities, and as the pathologies of racism and settler-colonialism continue to intensify, MSJ offers a transitional and transformational theory designed to address intersecting, relational more-than-human calamities and injustices.

Multispecies justice is simultaneously an analytic tool for understanding the current polycrisis and a normative framework for the development of transition, resilience, adaptation and transformative strategies, and for addressing injustices to all beings. It addresses as a matter of justice the flourishing of each element alone and *in relation*. By attending to radical co-mingling of all elements of the Earth system, the MSJ framework is not simply about attentiveness to wrongs against the more-than-human, but is key to the Earth system's protection, human and ecological security, and the protection of rights. Multispecies justice proposes an ethical and political frame that, we think, is more appropriate to the challenges of this era and the unfolding ecological and social crises that define it.

1.5 Section Outline

The sections that follow begin with a theoretical discussion and then move to a series of democratic experiments, examples, and proposals at multiple scales. Section 2 sets out some core principles for institutionalising MSJ, starting with the centrality of multispecies reflexivity, representation, and presence. We offer a critique of conceptions of justice that simply attempt to extend current notions in a limited way, in particular to some animals most like human beings. Building on long-standing work on Indigenous philosophies, critical theories from the Global South, and political ecology, we ground conceptions of ecological reflexivity, democracy, and inclusion in the relational ontology just discussed.

Sections 3–5 discuss the institutionalisation of MSJ in a variety of democratic experiments, most already in practice and some imagined and speculative. Some of our examples are from the Global South and Indigenous practice, but many are from the Global North because our aim is to explore how MSJ can be contextualised in a range of ways attentive to differences in place, culture, psychology, the relations present, and existing institutional realities. Section 3 focuses on institutionalisation and governance at local scales. It explores instances where local governments and communities are instigating practices of care, inclusion, and reflexivity for particular species and broader ecosystems. These include innovations in deliberative practice, practices of sustainable food systems, and designs for multispecies urban transition. Section 4 focuses on laws and courts, starting with the proliferation and pluralisation of novel legal rights and extensions of rights and personhood, often drawing on Indigenous worldviews and political practices. More speculatively, we discuss innovations such as wild law judgments and rights of nature tribunals, which re-read and redesign the (western) law for multispecies realities. Section 5 moves to the international and planetary levels, complementing existing innovations in Earth System Governance literature with broader and more speculative proposals informed by MSJ, including models of planetary institutions attentive to multispecies relationality and reflexivity.

We close with some reflections on the work that needs to be done to both repair and redesign political thinking and practice. How do people and institutions, with care and reflexivity, move beyond the impossible world imagined and implemented by liberalism, a world in which human beings make invisible, dominate, extract, decimate, and ultimately undermine Earth systems and relations? What are the next steps towards thinking MSJ into practice, living, and governing it with care?

2 Multispecies Justice from Theory to Practice

Moving on from the challenging theoretical innovations of MSJ as a set of ideas and critiques of dominant western notions of justice, this section examines how a revised conception of the subjects of justice can be embodied and implemented in institutional practices. In short, our definition of MSJ requires a theory of multispecies representation and presence, one awake to the breadth and heterogeneity of the multispecies communities included in MSJ. Multispecies inclusion stands as a response not only to the existing gatekeeping and silencing of the more-than-human in hegemonic human institutional decision-making but also as a direct rejoinder and response to the violence unleashed on those populations and the humans immersed within them. This violence, perpetrated by corporations, the state, and global political structures, is a direct result of philosophical and moral exclusion and invisibility. Multispecies justice depends on the visibility and presence of multispecies subjects and relationships in democratic decision-making – the institutionalisation of an ethic of ecological reflexivity. Institutionalising MSJ requires the design of inclusive, reflexive institutions capable of ongoing transformation where the more-than-human becomes a regular and influential component.

Here, we set out basic principles for institutionalising MSJ, building from the relational ontology laid out in Section 1 and bringing that into a democratic practice that centres multispecies reflexivity, representation, and presence. In contrast to more paternalistic extensionist forms of liberal representation, our approach embraces a form of ecological reflexivity engaged with the reality of decision-making in practice in different socio-ecological contexts. The project of institutionalising MSJ demands more than simply including the more-than-human in existing western forms of governance; it requires an exploration of what it means to have institutions embrace a form of relationality and radical permeability to the more-than-human, and how to afford reparative justice for those traditionally excluded and exploited – not just human alone but more-than-human more generally. Institutionalising MSJ is about re-embedding institutions grounded in place-as-relationships, with the reflexivity and responsibilities towards a growing array of subjects that deserve and demand recognition.

2.1 Limiting Justice and the Limits of the Extensionist Model

Restricting participation to those deemed to qualify as 'fully human' has, since Aristotle, been constitutive of the definition of the sphere of politics and of the category of subjects of justice in western thought. To lend legitimacy to such exclusionary criteria, the putatively exceptional human capacity for *speech* is singled out as the sole medium for distinguishing those deserving of justice and those not. In this way, speaking humans, and no others, are deemed to possess the types of rational and ethical sensibilities considered necessary for the political activity of collectively discussing and deciding on the best way to live together.[9]

Famously, Rawls made the argument that non-human animals are incapable of entering into contracts, specifically because of this lack of speech and rational

---

[9] '[S]peech serves to reveal the advantageous and the harmful and hence also the just and unjust. For it is peculiar to man as compared to the other animals that he alone has a perception of good and bad and just and unjust and other things of this sort; and partnership in these things is what makes a household and a city' (Aristotle 1905, 1253a). Note the term Aristotle used that is translated as 'speech' was *logos*, also translated as 'reason'.

understanding.[10] Hence, he argued, they could not offer the reciprocity necessary to become partners in a system of justice (Rawls 1971, p. 15). More recently, for Nussbaum (2007, 2023), the limiting factor has not been speech, but rather a sense of *dignity*, and a sense of an insult to that dignity – the understanding that one's life project has been interrupted by others. For Nussbaum, it is sentience, and not formal speech, that is the criterion for moral considerability; that sentience is also necessary for animals to have such a sense of dignity, and dignity is the entrée to consideration in a system of justice.

An ontological and epistemological Rubicon has thus been placed between this lauded suite of capacities – for speech, dignity, a sense of justice and ethics on one bank, and the possession of mere 'voice' on the other. The former provides access to moral and political personhood, while the latter can merely signal pleasure or pain, which do not, on this view, enable political inclusion or participation.[11] Staying with this schema, even more remote from the capacity to take part in politics are those who are considered 'mute'. Such non-subjects might be considered as objects of human considerations of proper action – right or wrong activities – but not as subjects of their own, with the interests, dignity, integrity, or reciprocity that opens the door to justice. In a relational world, in which the activity of a range of individual and collective human subjects is currently undermining environments, ecological processes, and planetary climate, such a division is self-destructive.

Against this background, there is by now a long-standing argument that, particularly in the context of governing, such exclusionary logics and practices have grave impacts on the interests of Earth others and impede human survival. Exclusion from the development and implementation of political decisions that profoundly affect them, we argue, constitutes a foundational injustice. We know that the interests of Earth others are undeniably and deleteriously affected by, and subjected to (human), political decisions. If one accepts that the legitimating condition for democratic politics is respect for either the All Affected Interests Principle (AAIP), which holds that everyone who is affected by a decision should get a say in the making of that decision (Goodin 2016,

---

[10] Although we use the term animals to refer to animals other than humans in this Element, non-human is consistent with Rawls.

[11] However, the distinction may not be as absolute as this standard portrayal indicates. As Julia Kindt (2024) points out, across his broader writings, Aristotle does allow that other animals (ants, bees, wasps, and cranes) are social animals and he used the term *zoon politikon* to describe them. Digging deeper, one sees that *zoon politikon* is used in both the more inclusive sense, whereby other species with certain collective decision-making practices are included, and the narrower sense, which is reserved for humans. As we shall see, many of the contemporary challenges rest on the view that the more inclusive sense ought to be accepted.

p. 366 and see Goodin 1996), or, more restrictively, the All Subjected Principle, which holds that those subjected to laws ought to be involved in making them (see Dahl 1989, p. 122, see also Näsström 2011), one has to conclude that their exclusion structurally inscribes multispecies injustice.[12]

Some authors understand the impacts of such exclusions and have offered to extend current liberal conceptions of justice to a limited number of more sentient animals (Nussbaum 2023). Such approaches, however, are limited in their scope and impact, focused as they are on inclusion of only the most sentient, individual animals, or those closest in relation to human beings. This kind of liberal extensionism continues to ignore ecological realities and relationalities, as well as the impacts on the interests of a much broader set of subjects. Vast forms of life beyond the supposedly sentient remain excluded, and those parts of the natural world beyond the living – such as minerals, mountains, and waters – may require some level of ethical attention but do not, on these accounts, rise to the level of considerations of justice. The ecological, the relational, and the immersive experience of human life among others cannot be encompassed with such a limited conception and practice of liberal extensionism.

Beneath the political-normative arguments around inclusion and exclusion rests the prior contestation, already raised in Section 1, that the 'rubicon' described in Section 2.1 has been placed in a way that is neither ethically nor empirically justifiable. This view comes in a range of slightly different versions, all substantively crucial to our argument. In one, it is argued that it ought not be the capacity for *speech*, defined in a way that describes what is unique to human forms of communication, but rather having *interests* that justify inclusion in political decision-making.[13] This approach not only embraces the AAIP or the All Subjected Principle but also has long been the basis of arguments for extending justice to Earth others in much environmental theory, such as that by Wenz (1988) and Baxter (2004). Baxter, in particular, extends such an understanding of affected interests not only to a range of animal subjects but also to the broader, collective reality of *species*. Here, we argue that addressing the interests of the impacts of the decisions and behaviours of (some) human beings and practices on the more-than-human should be encompassed by processes of justice.

---

[12] This operates across two dimensions, one being the procedural illegitimacy signalled by the AAIP, and the other the more substantive argument that excluding the more-than-human from taking part in political decision-making is more likely to result in neglect of their interests (see Magaña 2022). For a recent argument of animal participation critical of the All Affected interests approach and based on a recognition of human and animal interdependence, see Donaldson and Kymlicka 2023.

[13] We return to the meaning of interests and inclusion in Section 2.3.

In another view of inclusion, the argument is that the alignment of the qualification for political participation with the particular way in which humans know, organise, and communicate, rather than with the myriad other ways of knowing, organising, and communicating, betrays an unjustifiable arbitrariness motivated by an ideology of human exceptionalism rather than any reasoned justification. The growing empirical data on the complexity and speech-like character of the communication of numerous Earth others, flora and fauna alike, illustrates that, even accepting the asserted criteria, the boundary between subject and non-subject ought to encompass an ever-growing range of species and ecological processes. The more we learn about the breadth of communicative tools, and even language, in an ever-broadening array of species and multispecies relations, the more dubious this claim of human exceptionalism in communicative abilities becomes (Shah 2023).

Finally, and crucially, is consideration of the philosophies, ontologies, and practices of First Nations Peoples, for whom the conditions of political participation do not involve a separation between humans and Earth others. Respecting these demonstrates both the parochial nature of what is claimed to be a universal 'the way things are', and the colonial injustice to the multiple Peoples whose ways of thinking and living are consistently dismissed, undermined, or decimated. Without oversimplifying or universalising the Indigenous, the pattern across those cultures classified as Indigenous is to include everything in structures of relations. The earth, landforms, waters, air, elements, fish, fowl, and animal (including humans) are understood to live relationally, and thence human responsibilities and duties extend into all domains to maintain the integrity of the whole (Alfred 1999; Bawaka Country et al. 2013). Specifically place-based, the responsibilities are temporally (across generations from past to distant future) and subjectively expansive. In short, these are more realistic understandings of the Earth system and interbeing relationality than the dominating anthropocentric political and institutional imaginary.

On the basis of these contestations against a barrier between the characteristics of humans and Earth others, we argue that the process of institutionalising MSJ must involve finding ways of including the more-than-human in political decision-making. There must be a particular focus on those human communities whose more ecologically inclusive systems of governance were actively targeted for dissolution by settler-colonial forces. While it is crucial that considerations of justice – and injustice – remain focused on human behaviours that undermine the dignity, integrity, and/or life projects of a range of subjects, the point of a MSJ approach is to break open the barrier to the consideration of all of those impacted by human political, social, cultural, and economic decisions, and all of those deserving of functioning lives and futures.

2.2 Ecological Reflexivity and Ecological Democracy

Such consideration is the essence of the idea of ecological reflexivity. Institutionalised ecological reflexivity is a direct response to the damage done by western institutions that operate under the pretence that they are nowhere, or anywhere, and so they are free to despoil without sufficient consideration of the impacts on a broad range of subjects. The imaginary within which capitalism and the western state operates is one in which they, as entities, or the impacts they produce, are not attached to a particular place, with particular effects.[14] Such institutions are not informed by any reflexive ecological knowledge of place, nor by the cultural attentiveness of those who are – in fact, they are designed to ignore them, and they succeed only without such reflexivity. But if it was not apparent to those producing environmental impacts previously, on a planet in peril, it is now obvious that no institution is placeless. The imaginary and institutions of the liberal state are based on a radical departure from 'nature'. Institutionalising MSJ is about re-embedding and re-*placing* economic, social, and political institutions as a direct counter to the imaginary of nowhere. It is a direct counter to the settler state, a neocolonial corporation that has always been about control, extraction, exploitation, and territory, and not grounded in or responsible to place. Reflexivity and responsibilities come with this recognition.

A notion and practice of ecological receptivity is based in a pluralist ethos of engagement across difference, espoused by pluralist thinkers from William James (1977) to William Connolly (2005), and pluriversal thinkers from Arturo Escobar (2020) to a range of Indigenous scholars (Graham 1999; Whyte 2013; de la Cadena 2015; Watene 2016; Kimmerer 2013; Parsons et al. 2021; Winter 2022a). For Connolly (2005, p. 4), for example, a critical pluralist orientation requires a receptivity towards others, the ability to understand a variety of competing perspectives, and a critical responsiveness to this difference. Within the environmental justice movement, this kind of reflexive engagement across difference has long been key to creating a unified movement without uniformity (Schlosberg 1999).

In the recent canon of social theory, Ulrich Beck (1992) is well-known for his argument for what he calls 'reflexive modernisation', which is the process of coming to terms with the impacts and repercussions of the 'progress' of industrial society via regaining control of modernisation. For Beck, re-evaluating and taking back democratic and social control over an undefined notion of 'progress' is the necessary response to the creation of an increasing number of risks

---

[14] Worse, sometimes the identification of place is mainly used strategically to take advantage of lax tax, environmental and labour regulations.

in the social realm – what he calls 'the risk society'. A number of environmental political theorists have brought such an approach to reflexive governance together with a more thorough engagement with ecological realities. A particularly *ecological reflexivity* is an approach to representation and presence that begins with putting reflexive decision-making in practice within the context of numerous ecological issues and risks and the variety of cultural approaches and landscapes in which they are experienced (Schlosberg 2007; Dryzek & Pickering 2018).

Val Plumwood (2002) discussed this kind of ecological reflexivity as an attempt to bring into consideration the realities of those 'remote' from the usual political and economic decision-making, in particular those humans and Earth others bearing the worst ecological consequences of such processes. Similarly, Robyn Eckersley (2004, p. 115) sees ecological reflexivity as akin to a 'process by which we *learn* of our dependence on others (and the environment) and the process by which we learn to recognize and respect differently situated others (including Earth others and future generations)'. She understands this as a form of an 'enlarged mentality' that pushes imagination to include the more-than-human in our thinking.

Environmental and ecological democracy in political theory has long offered ecological innovations for broadening inclusion, representation, and/or participation of the more-than-human. This focus has been one key differentiation between more mainstream and pragmatic arguments for *environmental* democracy, which focuses on more representation of environmentally affected *human* communities in environmental policymaking, and *ecological* democracy, which acknowledges the desire for broader inclusion and institutionalisation of ecological entities and processes *themselves* (Pickering, Backstrand, & Schlosberg 2020). For Latour (2004, p. 58), the challenge of the latter is to a political theory generally, and a democratic theory in particular, that 'abruptly finds itself confronted with the obligation to *internalize* the environment that it had viewed up to now as another world'. This is the obligation of ecological reflexivity.

2.3 Multispecies Inclusion: Presence, Representation, and Participation

Turning from reflexivity to ecological and multispecies *inclusion* entails a number of still-unsettled difficulties, in three main respects. First, and continuous with debates about the scope of MSJ raised in Section 1, there remains significant disagreement about which Earth others ought to be included. In this regard, a key contention concerns the understanding of *interests* that informs judgements about who or what is capable of having them. Thus, some insist that sentience or subjective experience are necessary conditions, such that only those who can be *wronged* and not simply *harmed* are included (Frey 1983, pp. 154–155).[15] Others adopt a more open definition, where, as Ball puts it, 'x is in A's interest if x is necessary for and/or conducive to A's functioning or flourishing' (Ball 2006, p. 137; see also Eckersley 2011).

We argue for the more inclusive definition for two reasons. First, the relational ontology that undergirds MSJ renders incoherent the idea that one could attend to the interests of a select class of subjects of justice as if they were distinct from always being co-constituted through, and dependent for their functioning and flourishing on, those with whom they are in relationship, and whose interests thus also need to be included. In fact, from a MSJ perspective, it is not (or not only) the discrete interests of individuals or individual species that need to be included, but those of the relationships that will support functioning systems. Second, as discussed in Section 1, decolonising theories and practices of justice or, more positively, ensuring that the adopted theory of justice is articulated and institutionalised so as to include a broad range of (human) worldviews requires making space for those cultures for whom distinctions that have been central to western thought, like sentience or life, are not decisive, and perhaps not even meaningful. In short, limiting the interests of justice to individual humans, or even the most sentient animals, is both anti-ecological and colonising.

The second point of contention concerns the question of whether inclusion requires *participation* or can be satisfied by *representation*. In this regard, even those who recognise the deleterious impacts of this structural political exclusion and who advocate for democratic reforms to ensure inclusion of the more-than-human have, for the most part, insisted that the best that can be hoped for is a form of *representative* politics that will ensure that more-than-human interests are brought into political decision-making processes through what Anne Philips (1998) calls 'the politics of ideas', as distinct from 'the politics of presence'.[16] While acknowledging that 'the barriers to directly enfranchising those subjects are ones of practicality, not desirability', Robert Goodin (1996), for example, suggests that it is 'absurd' to think that Earth others could *actually* participate.[17]

---

[15] Alongside interest, another term used to draw distinctions is agency. For example, 'At its broadest, agency is the ability to have influence over, or have an effect on something. Agency in this sense is possessed by all humans and animals, but also by viruses, stones, or tornadoes (Carter & Charles 2013, p. 323). Our focus is narrower as we understand agency as the expression or manifestation of a *subjective* existence; agency means affecting the world in ways that reflect a subject's desires or will' (Blattner et al. 2020, p. 4).

[16] Relatedly, Dryzek (2002) would see such inclusion occurring through the inclusion of discourses representing more-than-human interests.

[17] For a discussion of this standard acceptance of representation with respect to animals, see Meijer 2019, p. 203 and Donaldson 2020, pp. 711–712.

Yet one should be wary about accepting judgements about what is practical at face value; they also inscribe some of the assumptions about the types of epistemic and communicative capacities required to participate in politics, and about the capacities Earth others possess.[18] Indeed, in light of MSJ theory's rejection of assimilative logics, and its condemnation of anthropomorphic criteria for being a subject of justice as a form of epistemic injustice, it would seem that institutionalising MSJ must at least entertain the possibility of a politics of participation, in some form. This is not to overlook the challenge posed, or the transformational nature of the reforms that will be required. The tremendous diversity of the modes of communication across the more-than-human world and their alterity from human speech mean that the forms of admissible political communication will have to radically expand in ways that would seem to break the frame of politics, so as to include diverse languages and embodied, gestural, material, and habitual forms of expression.[19] In the following sections, we explore some possibilities – existing, emergent, and speculative – about what these expanded forms and practices of 'political communication' might entail in practice.

In light of the undeniable practical barriers to such a thoroughgoing transformation, and how demanding it will be, our position is that, at this point, experiments in more-than-human participation should be encouraged, developed, and significantly amplified. In addition, existing institutional reforms will need to include an expanding range of representative practices.

Yet representation of the more-than-human is not without its problems, including the instrumental danger that human representatives will misrecognise others' interests or assimilate them into their own, and the more performative danger that insisting that only humans can actually participate reinforces the hierarchical ordering that is itself foundational to multispecies injustice.[20]

This then raises the third contention, which concerns the best *form* of representation. Given the paucity of experiments testing how different representative forms work in practice within modern nation-state politics, this remains a largely theoretical question. Nevertheless, there exists a large theoretical literature, for

---

[18] On assumptions about citizenships, see Donaldson and Kymlicka 2011, esp. 103; on political participation, see Donaldson 2020.

[19] This can be seen as an extension of Iris Marion Young's (1996, p. 124) argument that a 'communicative theory of democracy', including greeting, rhetoric, and storytelling, represents a necessary corrective to the exclusionary tendencies of the assumptions about appropriate forms of political speech in traditional theories and practices of deliberative democracy.

[20] Donaldson (2020) has an excellent discussion of this question in the context of domesticated animals, including of the antidemocratic nature of 'the competency contract' that this refusal of direct participation involves.

the most part divided between those exploring representation of 'nature' or the environment on the one hand, and those exploring representation for non-human animals on the other. These explorations canvas different forms of representation and evaluate them in terms of the various objectives and correlate dangers noted earlier in this section, such as accuracy or misinterpretation, legitimacy, abuse of power, paternalism, and more (see, for example, Ball 2006; Cochrane 2018; Donaldson & Kymlicka 2011; Donovan 2006; Eckersley 2011; Garner 2017; Goodin 1996; Hooley 2018; Meijer 2019; Parry 2016; Schlosberg 2007; Tanasescu 2014). Key questions concern the *who* and *how* of representation. Which humans are best placed to properly represent more-than-human others (ecologists, ethologists, environmental or animal advocates, those who live intimately with other animals or in a particular environment, including Indigenous Peoples)? How should their representative function be formalised (through trusts, delegation, reserved seats, ombudsmen, deliberative processes)? And how might representatives be held accountable to those they represent?[21]

Before discussing some examples of what political inclusion might look like (Section 3), it bears noting that, in practice, the distinction between representation and participation may be less absolute than it seems. In fact, when it comes to the practice of how more-than-human communication ought to be brought into political debate and decision-making, the classical dichotomy (participation/representation) may be an unhelpful distraction. For example, while rejecting the idea that Earth others can be given 'literal equality in the capacity to speak', John Dryzek (2002, pp. 153–154) nevertheless argues that democratic deliberation ought to include 'feedback signals emanating from natural systems'. Examples of such feedback signals could include rivers not flowing because of the (human) overuse or poisoning of water, species extinction due to habitat destruction, forests dying because of climate change, or soil erosion due to highly extractive agricultural practices. On an expansive view of political communication, these could all be understood as forms of participation, albeit ones that require attentive listening and careful interpretation. What is required is institutionalising ecological reflexivity. The role of humans, then, is not to represent the more-than-human but to attend to its communications, both perceptually and intellectually, so that they can be accurately interpreted and included. What needs to be institutionally factored in, however, are binding requirements that decision-making practices (and in some cases outcomes) take those communications, and the interests they convey, into account and that decision-makers can be held accountable when they fail to do so.

---

[21] Drawing from the disability field, Donaldson (2020) suggests the constitution of microboards, comprising a range of people who act as allies in the interest of the represented.

Indigenous movements, of course, have brought such issues of the lack of representation of *land*, of the more-than-human, to the fore in activism, court battles, and proposals for more place-engaged institutional design. Work in Indigenous environmental justice has long challenged the appropriation of land that lies at the heart of settler-colonial political institutions and explored the potential of multicultural and pluriversal allyship around the recognition and inclusion of the realities of, for example, water protection in institutional design (Whyte 2016). While western science has often been an ally of the institutional blindness to the more-than-human and an institutional ally of extractivism, we examine the extraordinary potential of sciences and traditional ways of knowing more aligned with MSJ, as recently detailed by Kimmerer (2020) and as embedded in approaches to innovative models of Aboriginal water management in Australia (Poelina et al. 2019), and listening to the world around us in Aotearoa/New Zealand (Ruru 2018). Such approaches illustrate the limitations of a simple and one-dimensional idea of 'personhood' and are instead attentive to less individualised and atomistic, and so more diverse, beings in the more-than-human realm, with designs for institutions that are grounded, representative, fluid, and cohesive.

2.4 Conclusion: Inclusion and Representation at Scale

In redefining institutions attentive to MSJ, the following sections introduce ways of thinking at a variety of scales, from the local to the planetary. This entails realisation of, and a response to, the sheer complexity of ecological reflexivity, which includes an attentiveness to a more-than-human world that is simultaneously microscopic and planetary. There are two crucial points to make in terms of scale here.

First, we embrace the polycentrism such reflexivity requires, given the intersecting nature of the more-than-human and the multiple forms of governance that flow from the grounded, fluid, and diverse realities of ecological systems. But in addition, and crucially, the institutionalisation of MSJ must be attentive to the need to avoid universalising any particular form or practice.

Second, polycentricity means an acceptance of a broad set of definitions of institutions, including the norms that frame and guide them, given the plurality of both ecological and cultural ways of knowing attached to places. While we aim to design MSJ institutions as decolonial, disruptive, and inclusive, such designs cannot assume a universalising frame. Institutions for governing more-than-human relations in place will be as diverse and as pluriversal as the range of cultures and places in which they will be applied.

While the goal is an institutionalised MSJ, flexibility and difference are crucial features informed by ecological reflexivity and a range of local experiences and knowledges. Such institutions still mean radical change that disempowers destructive, disrespectful, and unjust forms of governance and relations. Again, and crucially, MSJ is a critical theory, involving both critical/deconstructive and more transformative/constructive aspects.

3 Local Examples of Multispecies Justice

We turn now to some institutional reforms, starting with more obvious examples from the environmental democracy literature of deliberative assemblies and practices. We then examine some more substantive, prefigurative, and materialist forms of practice to illustrate the potential breadth of possibilities of more-than-human inclusion at the level of the local and everyday. Although, as noted throughout Section 2, it is Indigenous Peoples who have long adopted political practices that include the more-than-human, in this section our principal focus is on experiments in western contexts. We highlight these western innovations because, as set out in Section 1, it is western institutions that have been the source of the most systematic and the gravest multispecies injustices; therefore, it is western institutions that need to be transformed. In undergoing transformations, institutions may take inspiration from Indigenous philosophies, but these transformations will have to be appropriate to each institution's cultural context, to the relationships at play, and to existing institutional realities. Institutions neither can nor should imagine that Indigenous practices can simply be transferred (outside of culture and context) and applied to western-based institutions.

3.1 Deliberative Practices

In environmental politics particularly, there has been a long-standing argument that, even if they do not directly include representation or participation of the more-than-human, deliberative practices in themselves may be more likely (though not guaranteed) to develop decisions in the interest of the more-than-human. Goodin (2007) and Dryzek (2002), for example, both point to certain features of deliberative practices that are procedurally (rather than substantively) likely to increase attention to more-than-human interests, particularly where they take place in the context of evident ecological destruction. These features include the ways deliberation enhances critical reflexivity, demands accountability to others, broadens the range of interests included, and challenges entrenched hierarchies of systemic marginalisation. Given the recent growth of climate assemblies and the obvious and grave threats climate change poses to the more-than-human, one might see these forms of deliberation as an interim measure on a path towards more-than-human political inclusion. To find out whether deliberative processes can, in fact, give greater prominence to the interests of the more-than-human is, however, an empirical question, which ought to be subject to rigorous analyses and evaluations.[22]

A more fulsome institutional transformation would require more explicit attempts to include the more-than-human. In view of the potential advantages of deliberative forums, we would see them as a site for further institutional experimentation, although, given concerns about the overly 'rationalistic' processes often favoured, full more-than-human inclusion would require significant modification. At the simplest level, this could take the form of the proxy representative practices canvassed in Section 2, whereby nominated (human) participants are mandated to explicitly speak for more-than-human interests, or for specific interests relevant to the particular assembly (e.g. a threatened forest or species). A variation would be to draw in expert knowledges (including the knowledge of Indigenous Peoples or others with close knowledge of place) derived from the study of the communications of the more-than-human, for example, signals of the health, history, and trajectory of an ecological system. In this context, the rapid rise of the application of artificial intelligence to study animal communication is producing extensive data that could be deployed to assist (Rutz 2023). A more radical option would be to more directly bring in the communication of the more-than-human, for example by providing the human participants with an immersive experience in particular places and among particular ecological systems or animals – something that, again, new digital technologies might be well-placed to assist with. And yet, for all the theorising of such representation in deliberation, there are very few examples yet in practice of more-than-human inclusion in deliberative mini-publics in modern democracies.

Nevertheless, still marginal and emergent practices merit consideration insofar as they can be considered less formal and more prefigurative, where prefigurative politics are understood as '[f]orward looking, yet resolutely present' and work by activating 'imagination while reconfiguring lived social relations and the exercise of power' (Brisette 2016). Although often occupying peripheral spaces or taking place at a small scale, prefigurative politics can play an important role in denaturalising hegemonic assumptions and forms, and in

---

[22] As climate assemblies multiply, so does research on them. Duvic-Paoli (2022, p. 245) notes that Climate Assembly UK 'paid special attention to the objective of "protecting and restoring the natural world," which it ranked fourth in terms of priorities to achieve net-zero emissions ... [and] ... considered the environmental and biodiversity impacts of energy sources' and that in France, 'the climate assembly reinterpreted its emissions-focused mandate to include biodiversity protection, irrespective of whether it could contribute to reducing GHG emissions'.

affirming, in concrete practices and experiences in the present, that alternatives are indeed possible (and often already more widespread at local scales than many of us know) (Dinerstein 2016). Here, we consider three, each operating within modern liberal democracies and the very institutions we are suggesting must change for MSJ to be realised. They are intentional multispecies communities, experimental 'parliaments', and artistic experiments in multispecies world-making. Each offers innovative directions in deliberative inclusion.

With a conscious awareness of its role as a form of prefigurative politics,[23] VINE sanctuary in Vermont, the United States, describes itself as 'an LGBTQ-led farmed animal sanctuary that works for social and environmental justice as well as for animal liberation' where 'hundreds of animals co-create our unique multi-species community' (VINE Sanctuary 2017). VINE is explicitly committed to intersectional social justice across more-than-humans worlds, as well as to ecological protection and restoration, including by reserving over half of the physical space of the sanctuary as a wildlife refuge.[24] In addition to its philosophical embrace of ecofeminism, VINE manifests and leads in the praxis of MSJ in a number of respects. Specifically, the sanctuary is committed to and has been actively experimenting in practices where the non-human animal[25] residents take part in deliberation and decision-making.[26]

Some of the suggestions we have made for how deliberative forums could include the more-than-human are already practised by VINE. Donaldson and Kymlicka (2015) quote Pattrice Jones, one of VINE's cofounders, describing a type of immersive decision-making:

> We stood in the barn surrounded by sanctuary residents, as we like to do when making important decisions. [Sanctuary co-founder] Miriam and I have always believed that decisions about animals ought to be made, insofar as possible, in consultation with animals. If that's not possible, the next best thing is to be in physical proximity to animals like those you're thinking about, so that you don't make the mistake of treating them as abstractions. (Jones 2014, cited in Donaldson & Kymlicka 2015, p. 67)

---

[23] For example, one of its blogs says, 'those of us who want to live in a more just and peaceful world must begin to build that world in our own backyards' (VINE 2017 cited by Blattner et al. 2020).

[24] Apart from its active engagement with LGBTQI, racial, gender, and disability politics, 'becoming an engaged, empathetic member of the Springfield community is integral to the broader work of VINE Sanctuary'(VINE Sanctuary 2017).

[25] In Section 1, we indicated that we would use the term animal to refer to animals other than humans. In this context, we use non-human animals for clarity.

[26] Our analysis relies here on VINE's public material but draws extensively on the ethnographic work and analysis documented in Donaldson and Kymlicka (2015) and Blattner et al. (2020).

Blattner et al. (2020) and Donaldson and Kymlicka (2015) draw out several ways the VINE community operates to provide a supportive and responsive environment where non-human animals' individual and collective agency and decision-making can flourish, such that 'they are co-creators of their community' (Blattner et al. 2020, p. 17). Rather than particular individuals or species being placed in specified spaces, there is flexibility and ongoing negotiation over spatial and relational arrangements. The animals' expressed preferences influence where and with whom they live, with humans' roles being to attend and respond to those preferences. Routines and practices are similarly shaped so that 'negotiability of these decisions, not just the outcome, is an important dimension of agency – allowing individuals to be heard, acknowledged, responded to, taken seriously and to have the possibility to change an outcome to their liking' (Blattner et al. 2020, p. 7). The regulation of social norms constitutes a further aspect of collective deliberation, a complex task given the range of different species and individuals with different, sometimes traumatic backgrounds. In most settings where humans live with domesticated animals, it is the humans who determine which norms are 'appropriate' and then impose conditions onto the animals they live with (through for example training or segregation) to ensure these norms are respected. Blattner, Donaldson, and Wilcox document how, at VINE, it is the non-human animals, in relationship with the humans, who negotiate, transmit, oversee, and change norms concerning belonging and toleration, carefulness with others, and contact.

These practices may seem a long way from what counts as political deliberation and decision-making, but if one analyses them against the criteria deemed to define democratic politics – such as capacities for collective deliberation, co-authoring social norms, and committing to them – one can see that all those criteria would seem to be at work here. To insist that this does not count as politics because the medium through which it occurs is not human language, or because of the apparent absence of abstract reasoning, is to impose an arbitrary and unjust restriction.

We turn now from multispecies communities to co-creative multispecies artistic practices. This may seem like an odd place to find more-than-human participatory politics, but in recent years such practices have also been sites for experimenting with the development of participatory praxes in the face of the types of anthropomorphic and rationalist constraints on politics and political communication described in Section 3.1 (Sarkissian 2005; Sachs Olsen 2019, 2022). Recent work in this space is particularly illuminating because of the possibilities artistic practices offer for going beyond rational and linguistic forms of communication and because artistic practices themselves can act as a form of knowledge acquisition about the more-than-human.[27]

Sachs Olsen (2022), for example, analyses 'The Parliament of Species', an experimental art event co-staged to explore how Earth others could be included in the development of a park in the Oslo Fjord. Influenced by Joanna Macy's 'Council of All Beings' and Bruno Latour's 'Parliament of Things', the (human) participants were invited to explore the site, connect with a stakeholder of another species, and then 'ask' them a series of questions about how they lived, what was important to them and what kinds of transformation they would like to see or not. The humans then returned to the parliament to perform their speculations on the worlds and preferences of the multispecies others. In analysing how the parliament played out, Sachs Olsen is careful not to claim that the perspectives of Earth others were accurately represented. Rather, the process was able to provoke in the human participants a sense of the partiality of their own perspectives, recognition of the presence and alterity of the others living there and an aliveness to the critical but neglected role that attentiveness to diverse forms of communication plays in ensuring deliberative processes do not exclude marginalised perspectives.

Not dissimilarly, staged at various venues in western Europe in 2021 by an imaginary organisation called IOFLE (the Interspecies Organisation for the Future of Life on Earth), 'G5 Interspecies' was a science-fiction performance created by Spanish artist Rocio Berenguer.[28] Playing with and mocking the state-centric G8 and G20 summits, the performance placed human participants on stage with 'representatives' of animal, vegetable, mineral, and technology, including a creature created by the artist known as The Bad Weeds.[29] Although a curated performance, it, too, sought to provoke a reimagining of what negotiations might become possible when they include the more-than-human, with one of its aims being the creation of the first cross-species rights treaty. Furthermore, while its starting point was the political inclusion of the more-than-human via their separate kingdoms, the performance included a critique of the typological regime on which it rested, with 'hybrids' staging a rebellion and insisting on the impossibility of separating out species or living beings from the webs of continuity within which they exist.

---

[27] In discussing the development of what she calls Arts-Based Perceptual Ecology as a set of artistic methods for accessing the communication of the more-than-human world, Lee Ann Woolery (2017) discusses how 'scientists/artists such as Goethe (1749–1832), Darwin (1809–1882) and Haeckel (1834–1919) relied heavily on visual communication to explain their discoveries'.

[28] As part of its (fictional) identity, IOFLE was also responsible for WAFF, The wearefoodfoundation, concerned with regulating laws for 'who eats who'. See www.vivant2020.com/event/g5/?lang=en

[29] For an excerpt see www.youtube.com/watch?v=_mLaVtUoukk&t=139s

'La Démarche du Parlement de Loire' represents a more ambitious project combining different types of research, art, performance, public engagement, and public works. These were threaded together with a view to working out what a 'parliament' that included the entire more-than-human community that has a stake in or forms part of the future of the Loire River and environs might look like. It commenced with a series of hearings involving multidisciplinary researchers forming a type of commission on the form a parliament might take.[30] This was supplemented by days of study involving artists, urban planners, and water policymakers, exploring how water cycles and the culture of the river could be integrated into 'composing the territory', as well as a set of observational and immersive experiences of the river and the beings who live there, enabling more direct encounters with the river. This multidisciplinary research then formed the basis of an initial report entitled 'The river that wanted to write', synthesising the hearings and other research and setting out proposals for legal recognition of the Loire and the creation of an interspecies parliament (De Toledo 2021). Documentation included radio chronicles of the sensory environment, a documentary film, and archives documenting the genesis, development, and response to this early research. Other components of the project included a range of ecosystem and Earth art projects, such as sound maps, the development of a participatory game, and a range of architectural, artistic, and landscape projects across different municipalities. In addition, the project included support for urban and landscape planning projects mobilising cultural, anthropological, and ethnoecological approaches that were formed around and with the river.

All of these examples, from more formal deliberative forums to fictional performances to engaged learning, focused on a reflexivity and inclusion of the more-than-human in social and political decision-making. They are experiments in broadening existing democratic practices, and institutionalising attentiveness to the impacts of human activities.

3.2 Institutional Practice and Material Flows

Other forms of reflexivity and attentiveness focus more keenly on relationality and material flows, from bacteria in classrooms to the design of food systems in communities. As for the former, some of the most feared multispecies beings

---

[30] Virtually all documentation and commentary on the project is in French. See, for example, *Vers un parlement de Loire* (POLAU 2022) at https://drive.google.com/file/d/1zG74Bk9YH9WzO1F_Qa1ai 1V-I-2y3M24/view and 'Extraits de la revue de presse: La Démarche du Parlement de Loire' (POL AU 2021) at https://drive.google.com/file/d/1ceTjKEvnyG_Q05XTiVYWUALXcT6aCjO7/view.

are invisible to the naked eye. Microbes[31] stealthily pass from air and soil, water and cesspit, onto and into human and other animal bodies; they may assist their host to thrive or cause them to wither. It was their role as agents of disease and suffering that prompted the rise in antibiotics and antibacterial cleansers, invaluable tools for human health services over the twentieth century. Once-common bacterial infections are now less prevalent and potent because the science of deleterious microbes is well-advanced.[32] However, it is only more recently that the roles of 'good' microbes have been acknowledged, and that there has been growing recognition that their elimination (through wide use of antibiotics) may lead to poor health outcomes (Yong 2016) and a diminution of public well-being.

As more women move beyond the confines of the home and child-rearing and into the paid workforce, early childhood care and education centres have become ubiquitous in 'modern' western communities. These centres are charged with the care of families' treasured children and grandchildren and society's future generations. However, in the drive to protect them, the introduction of widespread use of antibiotics, sanitised surfaces, and 'clean' playgrounds has not only reduced infections from 'bad microbes' – a good outcome – but also reduced the communities of beneficial microbes in children's guts and on their skin. In protecting vulnerable children from harmful bacteria, their communities of beneficial internal and external microbes have been reduced, and their immune systems are weaker. This may have lead to problems like the growth in food allergies and some behavioural maladaptions (Puhakka et al. 2019; Roslund et al. 2020; Wang et al. 2022). Is this about MSJ? Yes. Multispecies justice is a *relational* theory: it seeks to protect the basket of relationships that allow systems and the individuals within systems to thrive. When we take the subject of justice to be relationships, then antibacterial interventions are unjust, and manifest in unexpected harms – such as a rise in childhood allergies. One intervention designed to enhance the microbial community within and upon children is the development of the 'microbial kindergarten'.

Early childhood centres are known sites of bacterial infections, but eliminating microbes from the childcare environment is not necessarily good for children's all-round good health. Recent experiments in Southern Finland explored the benefits that accrued following 'greening' daycare playgrounds.

---

[31] The term microbe is short for microorganisms, small organisms that cannot be seen with the naked eye. The term covers a range of organisms from different domains of life: bacteria, archaea, fungi, viruses, and some microscopic multicellular organisms like algae and protozoa.

[32] We equivocate here because it is unlikely that all microbes and their functionings have been identified.

Introducing forest floor material, sods, peat blocks, and planters into previously concrete- and gravel-covered preschool yards led to increases in the children's gut and skin microbial communities (Puhakka et al. 2019; Roslund et al. 2020; Wang et al. 2022). During outdoor playtime children ran, jumped, rolled, and climbed over these introduced natural and microbe-supporting new playground surfaces. Staff in the centres also introduced crafting activities using natural materials, and the children engaged in planter box gardening. The daily engagement with these natural playground elements brought a number of beneficial results (Puakka et al. 2019).

The children's play behaviours changed. Having the natural elements under play equipment and across the yard encouraged more adventurous behaviours, greater diversity of activities, the development of motor skills and coordination and increased physical activity overall (Puakka et al. 2019). The children enjoyed the nature-based playground: it 'offered embodied experiences of nature and provided the children with multi-sensory exploration and diverse learning opportunities' (Puakka et al. 2019, p. 1). A review of experiments demonstrated that exposure to a microbe-rich environment quickly enriched children's own microbial communities in comparison to children living and playing in nature-poor environments (Tischer et al. 2022). It also showed a positive correlation between these enriched microbiomes and the development of the children's immune systems and improved length and quality of sleep (Wang et al. 2022). This suggests the move to eliminate more-than-human relationships between children and microbes may be (in part) responsible for the rise in allergic and autoimmune disorders (Wang et al. 2022, p. 1). Reduced environmental biodiversity is linked to 'impoverished microbial exposures, [and the] consequent lack of promotion of homeostatic immunological development and ultimately the allergy epidemic' (Wang et al. 2022, p. 2). The push to eliminate bacteria, to exert human domination over the microbial realm, and to focus on only one aspect of human health has resulted in an imbalance to the systemic whole that leaves everything in some ways worse off: it has broken a fine web of relationality.

Microbial preschools reintroduce the reality of relationality, as well as a reflexive openness to the real consequences of such relations. In a similar way, sustainable food systems advocates have criticised the damage done by conventional agricultural systems that sterilise soil and take no heed of the flow of nutritional elements from soil, through plant, into human bodies and communities. The redesign of food systems practices thus represents another site for institutionalising ecologically reflexive democratic innovations. Here, there is an interesting overlap between reflexivity, multispecies inclusion, and the kind of material participation seen in the reconstruction and institutionalisation of such practices and movements.

The idea of material participation is about reworking embodied practices and processes, changing the everyday and ongoing flow of material things, such as food, as they flow through bodies, communities, and ecosystems (Schlosberg & Craven 2019). Material participation in food systems illustrates inclusive, participatory engagement with democratic practice – one focused on democratic control and oversight, a challenge to the power and injustice of the traditional food system, and a demand for sustainability through keen attention to the flow of materials through bodies and communities.

One key motivation for many in these movements is the desire to be reconnected to the full more-than-human world – to soil, to seasons, to ecological flows – in practice in the everyday life of communities. Activists working in food systems change repeatedly emphasise the importance of increasing individual and community engagement with, and involvement in changing, material flows. Such action is a matter of democratic action on the one hand, and the related ideas of social, environmental, ecological, and MSJ on the other. Food systems movements with such motivations include the growth of farmer markets, community-supported agriculture, local food policy councils, good food networks, and more. Such movements work to reshape the flows of food to address unsustainable and unhealthy food systems and to construct food systems that are good for farmers, eaters, and the ecological systems in which they are immersed (Alkon & Agyeman 2011; Winne 2011; Alkon & Guthman 2017). Ecological reflexivity is an everyday practice in these systems.

The focus of such movements is not just on individual material acts as politics – for example buying ethical produce as a form of individualist political consumerism (Micheletti & Stole 2012). Rather, ecologically reflexive material participation is about collective action that is attentive to social and environmental impacts of whole systems; it is about the mutual design and development of distinct and sustainable systems and a more transparent relationship between producers, consumers, and the broader more-than-human environment. Sustainable food systems movements are increasingly seen as alternative systems of the organisation and material flow of food through environments and communities across the globe, and they are expressed in a range of activities, from peasant movements of the south like La Via Campesina to food justice activism in California (Esteva & Prakash 1998; Petrini 2010; Sbicca 2018). Some of these materialist food movements are developed in response to food insecurity, others to preserve traditional foods and cultures, and many for ecological reasons, but most share the goal of challenging and changing undesirable, unsustainable, and unjust food systems.

The motivation for many is to physically remove communities from replicating the multiple social, ecological, and multispecies injustices of corporate agriculture and to develop new and just models, flows, and systems of the production, circulation, and consumption of food.

This is activism at a material level – participation in the redevelopment of a key flow of materials through bodies, communities, and environments. And it is activism with an ecological reflexivity at its centre; reconnecting to ecological flows is as crucial as reconnecting across communities via food. The sense among such activists is that 'just doing' can make positive change where policymaking and traditional lobbying and pressure fail (Schlosberg & Craven 2019). There is a sense that physically doing things with a view to changing a particular flow of matter – literally, in the case of food movements, getting one's hands dirty – is important, necessary, and political. Participants see this as legitimate democratic participatory action, with a goal of social and ecological justice, in a material and ecological form. Such material practice is a form of transformative ecological and multispecies justice and politics.

In this approach to material participation generally, and in the food example specifically, we see a crucial kind of social imaginary – a grounded imaginary (Celermajer 2021; Celermajer et al. 2024). The examples of activists on the ground reworking food systems – designing new and more sustainable flows of food through bodies, communities and environments – illustrate transformative practices, imaginaries, and just multispecies institutionalisation that is implemented and practised with the soil and with the local community.

The point here is that the kind of democratic and institutional innovations we are theorising, with an attentive and ecological reflexivity, a sense of the necessity of MSJ along with social justice, are already being practised. These are central examples of the potential for the provision of basic human needs – from childcare to food – with a keen attentiveness to more-than-human relationships and functioning.

3.3 Multispecies Cities and Architecture

Finally, the institutionalisation of MSJ can occur in ways that address the very makeup of everyday urban life. Cities epitomise the drive to monospecies existence: designed and built by and for (some) humans, they are often hostile environments for many Earth others. Urban design imposes human control over all Earth others, a talisman of human exceptionalism. With soil eliminated, or covered in asphalt and concrete, human feet encased in shoes forget ways to connect deeply with the earth. 'Wilderness' and the feared 'outsider' are banished beyond what are now mainly hypothetical city walls. Water, contained in and tamed by concrete and brick, is hidden, whisked from streets to places unknown. Parks and reserves are zones of careful curation and selected species, domestic yards covered in monocultures of mown grass. Animal pets are confined to the home or led through streets to the few green spaces where they are welcome (Meijer 2019), while animal 'enemies' – like rats, mice, or pigeons – are poisoned or trapped or repelled by spikes.[33] 'Weeds' – plants out of place – are sprayed or yanked from the earth. Bats and birds must range further or share crowded rooks and nooks. New Orleans is an exemplar.

New Orleans, a city 'saved from nature' by the superior intelligence of engineers (Gordon & Roudavski 2021, pp. 460–466) and city planners in the eighteenth to twentieth centuries, became the site of unnatural havoc during Hurricane Katrina, a storm itself the vile incarnation of fossil fuel extraction and burning. Had the river been left to meander as it had historically done, had the bald cypress trees still buffered land from bayou waters, the damage could have been better contained (Gordon & Roudavski 2021, pp. 460–466). Far from recognising how the original development had exacerbated vulnerability for humans and Earth others, however, the response to the impact of the hurricane on the urban environment echoes the hubris of earlier engineering responses. Although swamps are to be restored, nourishment areas identified, and the aide of a buttressing foreshore of bald cypresses is sought, rather than recreating a multispecies habitat the objectives 'continue to prioritise human needs. The instrumental framing of nonhuman lives casts animals, plants and other organisms as powerless and dispensible' (Gordon & Roudavski 2021, p. 467). Furthermore, in their already always capture to 'the most powerful stakeholders', '[s]uch approaches do not examine politics, fail to engage with issues of justice and self-determination, and continue to presume the superiority of human knowledge' (Gordon & Roudavski 2021, p. 468). Because these rebuilding projects are for (some) humans, unless there is a community-wide ontological reformation, they may be once again undone; they are an extension of extractivist logic – using Earth others simply for narrow human ends. While the ecosystem of the swamp may expand and revert to something of its former diversity, this is not the purpose of the project. We include this example as a note of warning as sea level rise seeps into and/or floods the world's major coastal cities.

---

[33] Spikes are also used to repel the homeless from sleeping in 'nice neighbourhoods' where they disturb the controlled and cultured aesthetic of the city. Hostile architecture, also known as disciplinary and defensive architecture, attempts to excise the homeless from the conscience of the city, designed-out of places to sleep by spikes, un-sleep-able benches or night-time water sprinklers. Like bees, pigeons, rats, and mice, the non-conforming human is provoked to leave: the architecture makes them invisible. See, for instance, Bader (2020) and de Fine Licht (2020).

In challenging the notion of cities as purely the space of human sociality (and capitalist excess), Bruce Braun (2005) suggests a perceptual refiguration towards a 'more-than-human urban geography' that acknowledges the plethora of entangled relationships with the more-than-human present in city spaces. The question Braun's and subsequent studies raise is one of how to live *in sympatico* or in some sort of 'ethical-political' (2005, p. 635) harmony with a more-than-human citizenry. Braun identifies that citizenry ranges from the inanimate materials of which the city is composed, and astride which it sits, to the curated presence of plant life and those few animate lives who tolerate cities' hostility. He foretold what is unfolding: something more than simply a sustainable city (Sheikh et al. 2023), but rather far-sighted planning that resettles urban and peri-urban more-than-human on their own terms.

Urban environments need not repel multiple species. While for some planners the reintroduction of plants particularly is targeted at supporting or enhancing human existence – for instance in the multiple calls to regreen cities to protect the human population from heat or to draw on 'ecosystem services' to enhance the lives of city-dwellers (Pollastri et al. 2021). Others suggest that cities might shift from mono-species pollutors to 'becoming active agents in reversing climate change' (Farinea 2020, pp. 249–250). Advances in computer imaging and AI software promise new opportunities to reimagine, model, and create 'buildings and urban environments [as] scaffolds designed for hosting multispecies coexistence and collaboration, [while] living organisms become, alongside human beings, bio-citizens contributing to multi-level systems of exchange and collective intelligence' (Farinea 2020, p. 250). Working with, not against, the intelligence of many beings could spur a transition from the modern cityscape, ill-suited for most living things, to future cities that are 'inclusive space[s] that foster[] dynamic processes of exchange' (Farinea 2020, p. 250) and allow for both humans and Earth others to flourish.

3.4 Conclusion: Resistance as Multispecies Justice

In this section, we have covered a range of forms of multispecies politics practised at the local level. Multispecies justice here manifests as resistance to anthropocentric politics in the context of animal sanctuaries, childcare, cities, and community gardens; as activism in art and sustainable collectives; and as politics-done-otherwise in relation with seas, rivers, and fjords. These are the politics of the everyday, beyond the structures and institutions of domination enabled by national and municipal politics, institutions, and regulation. We chose these examples from a long list of possibilities for their diversity and to demonstrate that, while the everyday is imbricated in the politics of multispecies injustice, small and manageable changes can be harnessed at local level to lead the wider community and institutions towards a comprehensive institutionalisation of MSJ. It *is* possible to challenge regimes of species-based hierarchies and instigate political acts of MSJ activism within our daily lives – indeed, these 'experiments' have resulted in widespread interspecies flourishing. In each case, it is not the individual human or Earth other who is the subject of justice, but rather the sets of relationships that bind the whole. The key takeaway is that, contrary to the mantra of capitalism and individualism, an individual cannot flourish unless there is space for 'others' to flourish: cooperation, not competition, with the more-than-human can lead to just, flourishing communities.

Individual or localised community actions to institutionalise MSJ are commendable and shine a path for others; however, they are insufficient for facing the urgency of the polycrisis. Put bluntly, the well-being of the planet depends on a rapid and radical reorientation from the politics of extraction, domination, and oppression. Fortunately, as we highlight in the next two sections, change at the constitutional and legislative levels is rippling around the globe.

4 Multispecies Justice and Law?

The past twenty years have witnessed significant innovations in and beyond the space of law designed to address escalating biodiversity losses and intensifying damage from climate change; these can be viewed as aspects of institutionalising MSJ. In this section, we consider some of the most significant of these innovations, before turning to the endeavours of citizens and activists to extend such advances in more radical directions. As climate-induced fires and floods advance (Christoff 2023) and more and more planetary boundaries are transgressed (see Section 5), such groups are emboldened to imagine new legal grammars and practices of democratic deliberation that better attend to the needs of multispecies communities. Given the extent to which law, a historically conservative institution, is routinely implicated in multispecies injustices across many domains, growing numbers of activists, lawyers and communities are experimenting with novel practices of care, imagination, responsibility, and intergenerational justice that extend far beyond the narrow confines of western modelled legal systems. We argue that these need to be carefully developed and amplified over the coming years, with pressure being put on the law to become much more responsive to the diverse needs of multispecies communities. This pressure will need to be exerted from multiple entry points, using many different strategies and doctrines, and in many different fora.

Efforts to move the law in the direction of greater multispecies justice might include more frontal attacks on the often incoherent and violent architecture of international and domestic environmental law that prioritises property rights and trade and investment priorities for capitalist corporations. They also certainly require a greater attentiveness to the diverse meanings and purposes of law in both Indigenous and non-Indigenous communities and a commitment to undoing colonial injustices that have become sedimented in a highly unequal world system (Ferdinand 2020). As Black and Indigenous scholars have often pointed out, MSJ demands an active programme of decolonising the law at many different levels, though exactly what that looks like in different national contexts is beyond the scope of this section. Imagining and enacting these practices (or some combination of these practices) is a prerequisite for the undoing of legal anthropocentrism and embedding of principles of MSJ into laws and legal institutions.

4.1 Legal Frameworks for Multispecies Justice: Rights for Nature?

Legal frameworks, both statutory and common law, have been slow to respond to theoretical developments in the area of ecocentrism. In the fourth edition of her seminal work on legal theory, published in 2017, Australian legal philosopher Margaret Davies included for the first time a chapter on 'ecolegality', in which she reflected: '[I]t is astonishing to think that nine chapters (and over twenty years) of a book on legal theory have pretty much managed to pass by without any mention, or at least not much, of the Earth and the ecological connections that make humanity and everything that flows from it possible' (Davies 2017, p. 451). This is indeed astonishing, given that the Rights of Nature movement and its offshoots, variously entitled Earth jurisprudence, wild law, and ecological jurisprudence, trace their origins to an article published forty-five years earlier: American law professor Christopher Stone's 'Should Trees Have Standing?' (Stone 1972). Stone's proposition, novel to the Euro-American legal tradition, that Earth others could have legal standing, enforceable rights, and the capacity to be represented in legal proceedings by human guardians was almost immediately endorsed in a judgment, albeit a dissenting judgment, by Justice Douglas of the US Supreme Court in 1972 (*Sierra Club* v *Morton*, p. 741). In fact, the publication of Stone's article was strategically timed to influence Justice Douglas's reasoning in this landmark case (Stone 2010).

However, it is only in the last fifteen years or so that, within the confines of western modelled courtrooms, rights of Earth others have been recognised in a sequence of decisions. These include recognition of the rights of the Atrato River by the Colombian Constitutional Court in 2016 (Lyons 2022), acknowledgement of the Ganges and Yamuna Rivers as living beings by the High Court of Justice in the State of Uttarakhand in 2017 (O'Donnell & Talbot-Jones 2018), conferral of legal personhood upon the rivers of Bangladesh by the Bangladeshi High Court in 2019, and recognition of the legal status and rights of the Colombian Amazon rainforest by the Colombian Supreme Court in the so-called *Future Generations* case (*Future Generations* v *Ministry of Environment*) in 2018. These represent only a fraction of the numerous experiments with more-than-human rights and Rights of Nature in recent years and many more are underway, even in places like Europe that have been slower to embrace this 'legal revolution' (Boyd 2017).

Such cases have initiated important conversations – as well as significant state and corporate pushback – about which sorts of rights can legitimately, usefully, and scientifically be attributed to rivers and other 'natural entities', with courts determining the decision-making authority of local communities (including the notion of 'biocultural rights'), the conditions under which previously granted permits for extractive operations in national forests can be revoked, and how to use ecosystem science to determine when specific 'rights' can be said to have been infringed (Kauffman & Martin 2023). Even in places where efforts to advance rights for rivers, lakes, and other ecosystems have been least successful at the national level, such as the United States, these rights have been viewed as threatening enough by the oil and gas industries to warrant the passage of state bills outlawing local efforts to adopt Rights of Nature provisions into municipal codes (Fitz-Henry 2024). Furthermore, they have ignited important debates about state pre-emption of local decision-making and contributed to growing awareness of the problematically close alliances between industry and regulatory bodies – alliances that are making the prosecution of MSJ nearly impossible in many, if not most, jurisdictions. Such conversations are critical to the practices of ecological reflexivity explored in Section 2, since communities at sub-national and regional scales will need to be empowered to engage in robust deliberation with and about Earth others with whom they are most closely entangled.

In the United States, a particularly important development in recent years has been the use of Rights of Nature to bolster Indigenous sovereignty and the many multispecies relations embedded in and enabled by that sovereignty. Early critics of the Rights of Nature argued that they were not necessarily aligned with Indigenous values (which were simply being appropriated or 'strategically essentialized' by non-Indigenous environmentalists) and that, as Australian Indigenous scholar Virginia Marshall has put it, these rights merely extend the colonialist impulse and may be used to further dispossess Indigenous communities (Tanasescu 2015; Marshall 2020). While these criticisms remain important – and there is much to consider about the limitations of rights discourses at a time of rampant corporate plunder of ecosystems and continued, often unapologetic settler-occupation of Indigenous lands – the last five years have seen important Indigenous-led efforts to use Rights of Nature arguments to protect lands and waters from oil pipelines and other large infrastructure projects. These efforts resonate strongly with those taking place in other settler-colonies, like Aotearoa/New Zealand, where Māori communities have played central roles in advocating for the creation of legal personhood for rivers and parks as acts of reparation from the Crown for colonial dispossession.

To provide just one example of Indigenous-led experimentation with the Rights of Nature from the United States: in 2021, the White Earth Band of Ojibwe in Minnesota filed *Manoomin et al.* v *Minnesota Department of Natural Resources* in the White Earth Tribal Court. The case aimed to protect the rights of wild rice, or *manoomin*, which is culturally and spiritually significant for Ojibwe communities. Like all rice species, *manoomin* is heavily water-dependent. The argument advanced by Ojibwe lawyers was that its very survival was threatened by the Canadian-based Enbridge company, which planned to extract large amounts of water for the construction of an oil pipeline – a move that had previously been approved by the Minnesota Department of Natural Resources. The case was eventually rejected because the jurisdiction of tribal courts extends only to the borders of federal reservations and does not apply to the decisions of non-tribal members that take place off-reservation. This, however, is in many ways an artificial distinction and one that a MSJ perspective would challenge – a distinction born out of hundreds of years of settler occupation that is inattentive to the ecological relations between reservations and surrounding ecosystems. Such decisions about limits on jurisdiction will need to be urgently revisited in the coming years to ensure that Indigenous Peoples are better able to enact MSJ without interference by settler-colonial governments or multinational corporations.

Despite this outcome, this was the first Rights of Nature enforcement case to be brought in a tribal court in the United States. Many more are likely to follow, as tribes develop innovative ways to use these rights to return to and strengthen long-neglected treaty rights. This is particularly important, we think, for the future of MSJ. As the authors of *The Red Deal: Indigenous Action to Save the Earth* point out, despite Indigenous Peoples making up just 5 per cent of the global population, 80 per cent of the remaining biodiversity in the world is on Indigenous lands (The Red Nation 2021). It is thus critical in the coming decades that Indigenous sovereignty – and particularly, the right to 'free, prior and informed consent' enshrined in Article 19 of the *United Nations Declaration on the Rights of*

Indigenous People – is honoured, deepened, and extended so that Indigenous Peoples, who often have millennia of ecological knowledge about particular places, can lead in discussions about MSJ and multispecies futures.

Despite these limited successes and creative experiments in North American tribal courts, movements for the Rights of Nature have arguably struggled in many contexts. After more than fifty years of the 'Great Acceleration', steadily rising emissions from the burning of fossil fuels and escalating biodiversity losses, the belated adoptions and adaptations of Rights of Nature in courtrooms are neither 'wild' enough nor sufficiently mainstream to halt the biodiversity crisis, which has been labelled (without hyperbole) the sixth mass extinction event (Kolbert 2014; Steffen 2015).

Statutory law has also achieved little in terms of MSJ, despite notable recent developments. These developments include, in Aotearoa/New Zealand, the path-breaking 2014 Act acknowledging the legal personhood of Te Urewera, previously a national park, and the 2017 Act granting legal personhood to Te Awa Tupua, the Whanganui River; a 2022 Spanish law recognising the legal personhood of Europe's largest saltwater lagoon, Mar Menor; and Panama's 2023 Rights of Nature law. Importantly, the *Te Urewera Act* and *Te Awa Tupua Act* put in place Māori guardians and are aligned with Māori cosmology. Yet such Acts are also limited and contradictory. For instance, section 64 of the *Te Urewera Act* continues to enable mineral exploration and extraction, and section 46 of the *Te Awa Tupua Act* excludes the water of the river, permitting the continued controversial operation of hydropower schemes that have serious adverse impacts on the health of the river and its ecosystems and contravening Māori law (Lurgio 2019). Furthermore, these Acts are still small in number in comparison to the vast body of environmental legislation which, although purportedly intended to regulate damaging activities, routinely permits and facilitates habitat destruction.

This is apparent, for instance, in Australia, where activists have campaigned for decades for stringent legislative protections for native species in their 'natural' habitats. In 1991, the North East Forest Alliance launched a campaign to protect the 'veritable forest-dependent zoo' (*Corkill* v *Forestry Commission* (*NSW*): p. 160) of Chaelundi State forest from logging. The campaign, which encompassed both direct action in the form of blockades and a test case in the Land and Environment Court, culminated in the introduction of the first threatened species legislation, the *Endangered Fauna (Interim Protection) Act 1991* (*NSW*), in New South Wales. An anomalous feature of that Act, however, was the so-called licence to kill (section 92B). Licences to kill, in varying forms, have persisted in subsequent threatened species legislative regimes in New South Wales and elsewhere, with their most recent iteration as biodiversity offsets. Biodiversity offset schemes establish a market for biodiversity credits, where credits are viewed as fungible items and traded to enable land clearing in areas containing remaining communities of endangered species. Such laws continue to facilitate the extinction of threatened species through habitat destruction.

These anomalies are becoming even more pronounced as climate change intensifies. As eminent legal scholar Mary Cristina Wood has observed, '[e]nvironmental law helped deliver [the] ecological emergency [of runaway heating] to our planet's doorstep' (Wood 2021, p. xxxiii). Astonishingly, and not atypically for Anglo-Euro-American legal systems, climate change continues to fall outside the remit of Australia's main piece of federal environmental legislation: the *Environment Protection and Biodiversity Conservation Act 1999* (Cth) (EPBC Act). A 2023 challenge mounted by the Environment Council of Central Queensland, contesting the decision of the federal Minister of the Environment to disregard the climate impacts of two proposed coalmines in carrying out risk assessment under the EPBC Act, was unsuccessful at first instance and subsequently on appeal (*Environment Council of Central Queensland* v *Minister for the Environment and Water (No 2)*; *Environment Council of Central Queensland* v *Minister for the Environment and Water*). In 2024, the High Court refused the environmental group special leave to appeal. The outcome of the Living Wonders case, so-called in light of the threats coalmine expansions present for Australia's threatened species and places, highlights the manifest deficiencies of the EPBC Act in the context of a climate crisis – and specifically its failure to recognise the entangled nature of climate, environmental destruction, and more-than-human well-being and the insufficient protection that the legislation affords to Earth others. In fact, two appellate judges noted that 'the arguments on this appeal do underscore the ill-suitedness of the present legislative scheme of the EPBC Act to the assessment of environmental threats such as climate change and global warming and their impacts' on protected areas and species in Australia (*Environment Council of Central Queensland* v *Minister for the Environment and Water*, para. 140). As climate scientists have recognised for some time, if these and similar projects proceed, it is not only legal loopholes like the 'licence to kill' that will eradicate remaining populations of endangered species.

Constitutional law has also not proven to be particularly useful in this regard, notwithstanding the radical step undertaken by the Ecuadorian government in entrenching Rights of Nature in its Constitution in 2008 and the recent inclusion of Rights of Nature in the constitutions of a number of Mexican states. Articles 71–74 of the Ecuadorian Constitution lay out the right of nature to 'integral respect for its existence and for the maintenance and regeneration of its life cycles, structures, functions, and evolutionary processes', giving people the right to petition on behalf of nature and requiring that the government appropriately remedy violations of nature's rights. Opinions among legal scholars, political scientists, and anthropologists remain divided about what these constitutional rights have been able to achieve in terms of initiating a new paradigm for sustainable development. Craig Kauffman and Pamela Martin have noted that, since 2018, after more than a decade of sustained failure in the face of a hostile and corrupt judicial system, these rights have been repeatedly brought from the lower courts to the Constitutional Court. This court, which is now staffed by independent lawyers, some of whom have been trained in Earth jurisprudence, has recently chosen to hear cases that have allowed them to consider these rights in a more substantive fashion and to create precedents about how to weigh them against other human rights: rights to economic development, to private property, and so forth. One of the most important of these cases is the Los Cedros case, which was decided by the Constitutional Court in 2021 in favour of the Rights of Nature, effectively banning mining and other extractive industries in a highly biodiverse national forest and, importantly, cancelling all existing mining concessions and environmental and water permits (Prieto 2021).

Such developments represent a significant expansion of the Rights of Nature, with both the Constitutional Court and provincial courts recently deciding that, even in cases in which environmental procedures (including environmental impact assessments) have been appropriately followed, development projects can still be halted on the grounds that they are violations of the Rights of Nature, the rights of rivers or communities' collective rights (e.g. in the Piatua and Aquepi River cases (*Realpe Herrera v SENAGUA 2021*; *Cristian Rigoberto v GENEFRAN 2019*). As these cases show, courts have begun to override the decisions of national Ministries and the National Secretary of Water (SENAGUA) when they approve development plans that compromise a river's ability to flow and to sustain the vital ecosystems of which it is a part. In the case of the Aquepi River, the Constitutional Court even went on to flesh out directives on how to 'conceptualis[e] and measure RoN violations in water-related ecosystems' by describing the 'structure, functions, and evolutionary processes' of water and water-related ecosystems (Kauffman & Martin 2023, p. 383). Such decisions concretise new methodologies for measuring and defending the health of riverine ecosystems, as well as underscoring the serious limitations of state environmental permits and embodying new norms around sustainable development by recognising that development projects need to 'balance competing needs in an integrated manner' (Kauffman & Martin 2023, p. 388).

These cases have also inspired important civil society experimentations with other forms of resistance to the activities and industries most responsible for the violation of the Rights of Nature, including most notoriously the fossil fuel industry. For example, on 23 August 2023, a historic national referendum was held in Ecuador on whether to permanently ban oil drilling in the Yasuní National Park, one of the world's biodiversity hotspots which is also home to at least two 'uncontacted' Indigenous communities. The referendum was initiated by many of the same civil society groups, including the youth-led collective Yasunidos, which had been instrumental in introducing the idea of Rights of Nature in the 2008 Constitution. The ban was overwhelmingly supported by the Ecuadorian electorate, making Ecuador one of the first countries in the world to issue a permanent ban on oil drilling in a part of its territory – no small feat given its historically heavy reliance on oil exports. These rights have also been taken up at the international level by the United Nations' Harmony for Nature Working Group (see Section 5) and by the world's largest international conservation organisations, including the International Union for the Conservation of Nature (IUCN)[34]. In 2012, the IUCN passed Resolution 100 calling for a process to include the Rights of Nature as 'a fundamental and key element' of decision-making 'with regard to IUCN's plans, programmes and projects'.

More orthodox Constitutions that do not include rights for nature increasingly include duties to protect the environment and citizens' rights to a healthy environment, but these, too, have for the most part not resulted in significant habitat protection nor robust reductions in national emissions. One exception here was the use, by a group of young plaintiffs, of a constitutional right to a balanced and healthful ecology to invalidate timber licence agreements in the Philippines, in the well-known case of *Oposa* v *Factoran* (*Re Minors Oposa* v *Secretary of the Department of Environment and Natural Resources*). Significantly, constitutional rights to a healthy environment increasingly feature in climate litigation against governments, with successful outcomes in two recent lawsuits – *Held* v *Montana* in 2023, now on appeal, and the 2024 groundbreaking settlement of a case brought against the Hawai'i Department of Transportation by a group of young people. However, environmental rights and duties are not a universal feature of constitutions. Certain constitutional texts, such as the Australian Constitution and the Constitution of the United States, contain no requirements or obligations in relation to the environment or Earth others.

---

[34] The IUCN is a large hybrid organization of governments and civil society organisations devoted to understanding the status of the natural world and the measures needed to protect it. Its 'Red List of Threatened Species' was set up in 1964 and has evolved to become the world's most comprehensive information source on the global extinction risk status of animal, fungus, and plant species.

There are, nevertheless, possibilities for 'wild law' retellings of constitutional case law, even with recalcitrant and archaic texts (Rogers 2014). Speculative retellings highlight the flexibility of judicial interpretation, and the potential of common law, or judge-made law, in relation to MSJ. Such avenues may well be more productive of legal change in practice than attempts to create new statutory or constitutional frameworks, even where these endeavours are as innovative and creative as Stefano Mancusi's 'playful' Plant Constitution: an imaginary constitution 'written by plants, and in the place of plants' (Mancuso 2021, p.10). Judges generally have more independence from established interests and powerful lobby groups than do elected politicians, and hence they have more capacity to pursue just outcomes that may prove to be elusive or unpopular.

In the following section we address ways in which the common law can be prefigured or reimagined to embrace the principles of MSJ. In recent years, this challenging task has involved experimenting with the retelling and rewriting of existing judgments, from the perspectives of other species. It has also been undertaken through the work of various Rights of Nature tribunals. We briefly explore each in turn.

4.2 Wild Law Judgments and Rights of Nature Tribunals

The Wild Law Judgment project was launched by Nicole Rogers and Michelle Maloney in 2014, as one of a suite of critical judgment projects which encompass feminist rewritings, Indigenous rewritings, and most recently the interdisciplinary Anthropocene Judgments project. Such projects are experiments in creating new 'legal imaginaries' which, in turn, can shape orthodox legal forms and institutions (Grear 2020). They are efforts to imaginatively push the law towards a fuller attentiveness to the needs and interests of the more-than-human.

In the Wild Law Judgment project, participants adopted a more expansive methodology than that of existing feminist judgment projects; wild law judgments could include hypothetical judgments based on new forms of ecocentric law, as well as existing judgments rewritten from a wild law perspective. The pursuit of MSJ was a prominent aspect of the project. The current Chief Judge of the New South Wales Land and Environment Court, Brian Preston, contributed not only an essay on writing judgments wildly but also a futuristic judgment originally delivered at a mock trial in 2012, in which green sea turtles brought an action in public nuisance against the Commonwealth and Queensland governments for destroying their habitat in the Great Barrier Reef through the ongoing approval of coalmines in Queensland (Preston 2017). In other judgments, a group of Queensland air-breathing lungfish sought to protect their spawning sites by taking action in trespass against the operators of a dam (Coyne 2017), and the 1948 *International Convention for the Regulation of Whaling* was amended to create the position of a special Representative for Whales and to exclude current exemptions for whaling (Johnson, Lewis, & Maguire 2017).

In one of the more remarkable judgments in the project, the writer Bee Chen Goh revisited the celebrated English tort case of *Donoghue* v *Stevenson*, from which the jurisprudence of negligence derives. The facts of this case are well-known to all law students: a Scottish shop assistant accidentally consumed part of the decomposing body of a snail when drinking from a bottle of ginger beer in a café in 1928. The striking feature of the decision was the attribution of liability to a third party, the manufacturer of the ginger beer. In her rewriting, Goh, a practising Buddhist, focused upon the fate of the snail: did this hapless creature, as a sentient being, have any cause of legal action against those who contributed to its demise? After reflecting upon various tenets in Buddhist philosophy and noting that 'Buddhist thought is akin to wild law philosophy', she found for the snail (Goh & Round 2017, p. 100).

Subsequent critical judgment projects also encompass speculative judging with MSJ as a foremost consideration. Participants in the UK Earth Laws Judgment project, launched in 2019, have provided Earth Law retellings of key judgments in the United Kingdom (Dancer, Holligan, & Howe 2024). The Anthropocene Judgments project (Rogers & Maloney 2023), in which participants were tasked with creating the judgments of the future, contains a number of judgments oriented towards MSJ. Here, future judges prioritise protection of the habitat of endangered Tasmanian devils and revived Tasmanian tigers (Jessup & Parker 2023), delineate the rights of cow inhabitants of a smart dairy farm in 2057 (Szablewska & Mancini 2023), speak for the ocean and its Earth other inhabitants (Abedesi 2023), and emphasise the importance of open borders for Earth other climate refugees such as wild swans (Dao 2023). Michelle Lim envisages novel judgments delivered on a future Judgment Day by Matang Mountain, Kilimakyero Lichen and the Little Bush Warbler: the last delivered as a song in Chinese (Lim 2023).

Rights of Nature tribunals are another important mechanism for demonstrating possibilities for MSJ in the common law. These tribunals are, as Robert Cover put it when discussing the 1967 International War Crimes Tribunal established by Bertrand Russell and Jean-Paul Sartre, 'a philosopher's realization of an ideal type' (Cover 1985, p. 202). Tribunal findings lack what Derrida famously called the 'force of law' (Derrida 1990, p. 920). Such findings can, nevertheless, play an influential role in law reform.

The first International Tribunal for the Rights of Nature was established in 2014 at the Global Alliance for the Rights of Nature (GARN) Summit in Ecuador. Tribunal 'judges', including leading philosophers, scientists, lawyers, First Nations representatives, and activists, were guided by the principles in the non-legally binding *Universal Declaration for the Rights of Mother Earth* in reaching their conclusions on cases brought before them. This declaration, written by more than 30,000 civil society activists in 2010 in the highland city of Cochabamba, Bolivia, remains one of the most radical and far-reaching declarations of the urgent need for global climate justice anywhere in the world. It is also the first to include explicit recognition of the 'rights of Mother Earth' (Fitz-Henry & Klein 2024).

In one of the earliest tribunals in December 2014 in Lima, Peru – scheduled to coincide with COP20, which was held concurrently in the same city – cases were brought on behalf of the reptiles, insects, and plants of Yasuní National Park. These 'plaintiffs' were seeking protective action because they were being threatened by intensifying oil extraction as part of the Yasuní-ITT project that had just been approved by the Ecuadorian government (International Rights of Nature Tribunal (a)). Much to the outrage of environmentalist and Indigenous rights organisations and communities the world over, oil exploration had been given the green light in 2013 by President Rafael Correa after the failure of his multi-billion-dollar proposal for the international community to compensate Ecuador for leaving millions of gallons of crude oil underground. In his decision at the Tribunal in Lima, Ecuador's former Minister of Mines and Energy, Alberto Acosta, pointed out that if all the crude oil from the Yasuní-ITT block was to be extracted and burned, based on estimates of current global demand, it would last a mere nine days. However, it would mean the wholesale destruction of forest communities whose intense connectivity and global importance many human beings are just beginning to appreciate (Jabr 2020). Acosta argued that this was a clear violation of the Rights of Nature or, as he put it, referencing the Cochabamba People's Declaration of 2010, the 'rights of Mother Earth'.

Other cases also focused on the dangers posed to Earth others by extractive industries. In another, brought by the Ecuadorian ecological rights group, Acción Ecológica, the argument was that the rights of the marine ecosystems in the Gulf of Mexico had been violated by British Petroleum's (BP) Deep Horizon oil spill in April 2010. While BP had paid millions of dollars in damages to affected *human* communities along the Louisiana shore in the southern United States, the costs of properly restoring all the marine ecosystems affected would have been considerably higher. This was another case in which plaintiffs argued that MSJ had not been served because previous litigation had not fully considered the vast range of marine relations affected by the spill nor the extended temporalities required to make sense of the ongoing, intergenerational damage to those relations. To substantiate their arguments, the scientific team noted that more than 900 dolphins had died or been stranded; thousands of fish and other marine mammals were found to have abnormally elevated hormone levels, lung diseases, or anaemia; and at least 500 sea turtles had perished. In their final judgment, the Tribunal recommended, among other things, a moratorium on all deep-sea oil drilling (International Rights of Nature Tribunal (b)). They also issued a plea to the United Nations to create a collective multi-lateral process to assess petroleum operations, to consider and impose moratoria, and to identify necessary reparations for disasters past, present, and future. Such demands point the way towards what a more radically just and less anthropocentric future might look like and the sorts of legal fora that might steer us towards such a future – ideas that we explore further in Section 5.

Subsequent Peoples' Tribunals investigating violations of the Rights of Nature, both international and domestic, have been established in various jurisdictions. In Australia, the Australian Earth Laws Alliance created a permanent Australian Peoples' Tribunal for Community and Nature's Rights, which has held three public hearings and undertaken a number of inquiries. In one such inquiry, the Tribunal investigated the Black Summer megafires of 2019–20 and considered whether the systemic failure on the part of Australian governments to Care for Country constitutes ecocide. In another inquiry in 2016, the Martuwarra River was presented by Traditional Owners as an ancestral being deserving of rights – a being subject to 'First Laws' that long predate the precepts, institutions, and practices of settler-colonial law (Martuwarra RiverOfLife et al. 2021).

Tribunal proceedings and findings can have ramifications far beyond the borders of the law. It is worth noting, for example, that although such tribunals are often dismissed as little more than symbolic performances, nine years after the Yasuní judgment at the GARN Tribunal in Lima the Ecuadorian public voted on a referendum to ban oil drilling within the national park – a reality that seemed unthinkable in 2014. And such tribunals do other cultural work: making space for intercultural exchange about diverse legal traditions and the limitations, blindness, and violence of settler-colonial law; making room for the shared expression of grief about the loss of ecosystems and more-than-human communities; and incubating possibilities for more radical legal precepts to come (Fitz-Henry 2017).

Davina Cooper has speculated upon the significance of imaginative and performative initiatives such as judgment rewriting projects and Rights of

Nature tribunals. She describes these as a form of mimetic play, or 'State play with revisions' (Cooper 2019, p. 25), as alternative apparatuses for both imitating and reimagining state institutions. Such projects can be viewed as part of a 'creative refusal to give up and give way' in 'conditions where institutional formations ... seem radically inaccessible to progressive and radical forces' (Cooper 2017, pp. 187, 208). She believes that there are possibilities for 'public governance to take up and be shaped by new kinds of articulation' (Cooper 2019, p. 163). In 2018, Brian Preston, a sitting judge, observed that 'once we give the right-less thing rights or have considerateness for it, the unthinkable becomes thinkable ... The Wild Law Judgment Project has started this process of making the unthinkable become thinkable' (Preston 2018, p. 226). Imaginative exercises in viewing law through 'fresh eyes', whether through judgment rewriting, speculative judgement writing and/or peoples' tribunal hearings, are precursors to institutional change (Preston 2018, p. 226).

In the following section, while noting the multifaceted and multidirectional promise of these imaginative legal exercises and practices, we argue for even further widening and radicalising efforts to reorient legal systems towards MSJ.

4.3 How Can We Push These Imaginings and Practices Further?

Cultural phenomena outside these specific forms of extra-legal performance can also influence and reshape law to accommodate considerations of MSJ. Such phenomena include art, storytelling, and certain forms of activism – practices that we began to explore in Section 3. For example, artist Amy Balkin, in an artwork called 'Public Smog', seeks to protect the atmosphere, and all beings dependent upon a life-sustaining atmosphere, through the creation of an invisible and intangible public park in the sky and World Heritage listing of the atmosphere (Balkin 2015). This is art with distinctive practical application and projected legal outcomes. Legal scholar Michelle Lim has imagined the voices and stories of endlings, the last of a species, in the hope that 'endlings might form the catalyst for developing a more responsible, more empathetic future law' (Lim 2020, p. 614). First Nations scholar Irene Watson deploys story (Watson 1997, pp. 43–44) in her depiction of 'raw law': 'a natural system of obligations and benefits, flowing from an Aboriginal ontology' (Watson 2015, p. 1). Raw law is transmitted 'through living, singing and storytelling' (Watson 2015, p. 12). Watson questions the legality of approvals and licences emanating from colonial systems of environmental law, pointing out that '[t]here is no lawfull authority held to consent to destruction of the land, for that is the law' (Watson 2017, p. 210).

Embodied performance on the part of activists is another cultural mechanism for legal change. Forms or practices of activist performance can present compelling ideas of MSJ through what Kevin DeLuca identifies as 'image events' (DeLuca 1999). Examples of this could be seen at the protests that occurred during Australia's Black Summer – the repeated appearance of a giant, charred, partly skeletal koala, as well as clever imagery and wordplay by Brandalism street artists depicting literary icon koala Blinky Bill fleeing the fires, reinforced for bystanders and observers the daunting toll of the megafires on Earth others.

Activism also encompasses lawbreaking. A key purpose of civil disobedience is to instigate legal and institutional transformation through transgressive performance. For environmental activists, the changes they seek encompass MSJ. Yet the viability of environmental activism is increasingly undermined by the enactment of draconian laws designed to deter protesters.

The pro-growth bias that permeates current legal systems and the collusion of law enforcers in upholding and reinforcing the tenets of neoliberal capitalism are increasingly manifested in contemporary legal encounters between governments and non-violent defenders of nature. For instance, in New South Wales, the latter could face significant prison sentences and/or fines for disrupting a (non-sentient) road and the passage of human traffic (*Roads and Crimes Legislation Amendment Act*) until parts of the legislation were recently invalidated (*Kvelde v NSW*). Yet, as already highlighted, significant disruption of the living habitat of threatened species by corporate developers and forestry attracts minimal penalties if unlawful and, in many instances, is rendered entirely lawful by licensing and other permissive regimes. Similar anti-protest laws have been enacted in other jurisdictions, including the United Kingdom, resulting in jail terms for some climate activists (Gayle 2023). And of course Indigenous land and water defenders in the Global South have long been disproportionately subject to state surveillance and criminalisation for simply attempting to exercise sovereignty over their lands in the face of pipelines that threaten their water and water-dependent kin (Estes 2019). To provide just one of many examples, in 2014 a prominent Indigenous anti-mining organiser from the Shuar community of southern Ecuador, Jose Tendentza, was brutally killed just a few days before he was due to travel to Lima, Peru, for the International Rights of Nature Tribunal described earlier. Although the state was never held accountable for this murder, it is widely believed by the Shuar community and NGO observers that he was killed for his well-known opposition to one of the country's largest open-pit mines: the Mirador copper project.

In the face of this increasingly hostile legal and economic climate across the globe, some organisers and scholars have begun to call for even more radical efforts to transform legal systems. In his widely read and cited book *How to*

Blow Up a Pipeline, for example, Andreas Malm has argued for the importance of more violent direct action against fossil fuel infrastructure (and, we might add, against the legal system that supports and protects that infrastructure). Noting the limitations of non-violent civil disobedience, or what he calls 'strategic pacifism', Malm argues that we are well and truly out of time to save many of our fellow Earth-beings. Given the 'extraordinary inertia of the capitalist mode of production meeting the reactivity of the earth', there is a need, he suggests, for more robust debate about new activist tactics that might begin to dislodge the iron grip of the fossil fuel industry on legal regimes throughout the world – a grip that is making planetary chaos all the more likely and that is effectively sentencing millions of species to untold destruction (Malm 2021, p. 66). While we are not advancing an argument for violence, we are simply noting that the process of institutionalising MSJ in and through legal systems will not be straightforward and may, in fact, involve significant destabilisation before transformation.

4.4 Conclusion

In this section, we have briefly explored recent legal innovations that might ultimately deliver greater MSJ. Specifically, we have noted the limitations of recent experiments with Rights of Nature in the face of both mainstream environmental law, which continues to prioritise the protection of property rights, and international trade and investment law, which continues to uphold and defend the prerogatives of multinational corporations. However, we have also highlighted some of the ways in which these experiments have initiated broader conversations about Indigenous rights and the need to further decolonise the law, imaginative retellings of case law that allow for the development of more expansive understandings of responsibility towards the more-than-human world, and the articulation of symbolic judgments that might point the way towards the more egalitarian and less anthropocentric tribunals of the future. We have concluded with a demand for even more radical action – in art, activism, and law – and have suggested that the path ahead will likely involve exerting ever-greater pressure on existing legal systems through the advancement of novel rights claims while also engaging in civil disobedience when those systems continue to authorise widespread death and destruction of more-than-human communities.

5 Planetary Institutions for Multispecies Justice

We now turn to institutionalising MSJ at scale. Global environmental agreements are filled with strong statements of purpose that seem to be focused on ecological concerns: the 1992 *United Nations Framework Convention on Climate Change* (UNFCCC) aims to prevent 'dangerous anthropogenic interference in the climate system'; the 2015 *Paris Agreement* on climate change notes 'the importance of ensuring the integrity of all ecosystems, including oceans, and the protection of biodiversity, recognized by some cultures as Mother Earth'; and the 2022 *Kunming-Montréal Global Biodiversity Framework* 'aims to catalyze, enable and galvanize urgent and transformative action by Governments, subnational and local governments, and with the involvement of all of society to halt and reverse biodiversity loss' (United Nations 1992a, 2015; *Convention on Biological Diversity* 2022). However, as we look across treaty after treaty, they are all failing to achieve their goals and secure a habitable Earth for all its beings. Interstate powerplay and deadlock, corporate lobbying, and poor institutional design explain much about these failures. Yet there is a deeper failure of inclusion and membership at work. A MSJ lens invites us to see whose interests, lives, and agency are represented there, and whose are missing: the voices and flourishing of Earth others and human communities – especially Indigenous Peoples – who are most closely entangled with their worlds.

This section analyses the problem of globalising MSJ in two ways. Its first section considers how existing environmental governance marginalises the project of MSJ and explores modest reforms within existing treaty and international architectures; that is, within the limits of the *United Nations Charter* and major conventions. Its second part proposes more far-reaching reforms that we think would be more effective, democratic, and representative of ecosystems and the more-than-human at the international level. It draws inspiration from new scholarship in world politics and international law that foregrounds the presence and rights of nature and the entangled questions of justice that link human and more-than-human communities in ever more compelling ways (Pereira & Saramago 2020; Youatt 2020, 2023; Natarajan & Dehm 2022; Fishel 2023; Arvidsson & Jones 2024).

We present proposals for new institutional and diplomatic designs, but equally important is a profound shift in understanding of who and what governance is for and the worlds it inhabits and sustains. Because human political and legal languages are recent, specialised, and idiosyncratic inventions, the more-than-human will appear at one or more removes. Faithfully understanding and representing the interests of ecosystems (or the entire planet) will be a fraught responsibility that entails ongoing risks of failure and needs constant re-evaluation (and revaluation) in a dynamic Earth system. The fact that international forums are organised in a way that precludes the participation of Earth others does not mean that they do not speak or communicate information that is crucial there. We disagree with Aristotle, Habermas, and even Dipesh

Chakrabarty that Earth others cannot be political actors or 'subjects'; that, in Chakrabarty's words, 'the proactive question "What is to be done?" is still for humans alone' (Burke & Fishel 2020b; Chakrabarty 2023, p. 16).

At its root, MSJ challenges the ontological separation of human worlds and the worlds of Earth others, and it challenges us to bring forth more-than-human agencies, lives, and concerns. It reminds us that we all live in complex webs of relation and dependence across species and worlds. Governing for the planetary *is* an urgent question for humans, but we will never do so alone.

5.1 Reforming Anthropocentric Governance Regimes

Humans and Earth others live and entangle at multiple scales that extend from the microbial to the planetary. Yet global governance is primarily state-centric, and environmental governance is extraordinarily recent. The 1945 *Charter of the United Nations* makes no reference to the environment. Even the 1972 *Stockholm Declaration of the United Nations Conference on the Human Environment* is an anthropocentric document, concerned with the interest of 'the peoples of the world in the preservation and enhancement of the *human* environment' (Sand 2008).

5.1.1 Problematic Principles and Environmental Human Rights

Eco- and anthropocentric commitments in international law are encoded in principles which shape future law-making. These principles have contradictory commitments and effects, and even where more ecologically rigorous principles exist in international law, states routinely ignore them. Two of the principles set out at Stockholm – 21, permanent sovereignty over natural resources (PSNR), and 22, on the prevention of transboundary environmental harm – have become international customary law and give states 'the sovereign right to exploit their own resources pursuant to their own environmental policies' subject to weak obligations not to cause harm beyond their national jurisdictions. While the PSNR principle was framed to defend states' rights to control and regulate foreign investment and transfers of wealth to the already rich nations, it also licensed unrestrained mining, logging, and habitat destruction. Another key principle, 'Sustainable Development', which was the centrepiece of the United Nations Conference on Environment and Development (UNCED) in 1992, has barely been able to limit this destruction (Gordon 2015, p. 61).

A new principle also introduced at UNCED – the 'Precautionary Principle' – had a more ecocentric cast that could help governments grapple with a dynamic situation of Earth system crisis where multiple planetary boundaries are under stress or being exceeded. It states that 'where there are threats of serious or irreversible damage, lack of full scientific certainty shall not be used as a reason for postponing cost-effective measures to prevent environmental degradation' (*Rio Declaration on the Environment and Development* 1992, sec. 15). From a MSJ perspective, precaution should apply to the full range of planetary ecological relations of which humans are a part. Precaution values them all. The Precautionary Principle is now considered international customary law, but it has been increasingly marginalised in major treaties, absent from both the 2015 *Paris Agreement* and the 2021 *Glasgow Climate Pact*. Another principle introduced at UNCED 1992, 'Common But Differentiated Responsibility' (CBDR-RC) is, however, prominent, because key state consumers and exporters of fossil fuels from the BRICS[35] and non-western nations have weaponised it to delay emissions reductions and energy transformation (Burke 2022). At the same time, the principle highlights the historical responsibility of the United States and other developed states (including China) for climate pollution and damage.

Fifty years after Stockholm, environmental human rights were recognised in a new UN General Assembly resolution, 'The human right to a clean, healthy and sustainable environment' (United Nations General Assembly 2022b). The resolution notes that 'environmental degradation, climate change, biodiversity loss, desertification and unsustainable development constitute some of the most pressing and serious threats to the ability of present and future generations to effectively enjoy all human rights', but its concerns are limited to human beings. It makes no references to the other important General Assembly Agenda, 'Harmony with Nature', which is focused on the agency and intrinsic value of the more-than-human realm (Schmidt 2022).

5.1.2 Key Conventions: Climate Change and Biodiversity

While there are now thousands of multilateral environmental agreements, in all but a few cases they are failing to arrest the global ecological polycrisis as it manifests across climate change, biodiversity loss, pollution, deforestation, desertification, environmental racism, and the killings of environmental defenders. The 1992 *United Nations Framework Convention on Climate Change* was concluded as the Earth passed the safe 'planetary boundary' for CO_2 – 350 parts per million. Half the excess carbon dioxide in the atmosphere has been emitted

---

[35] The BRICS are a group of five large, 'emerging' economies – Brazil, Russia, India, China, and South Africa – that is slowly becoming more institutionalised and challenging aspects of the western-dominated economic order. They cover 27 per cent of the Earth's surface and account for nearly 30 per cent of global GDP and 45 per cent of annual global CO_2 pollution. In 2024, BRICS membership will expand to include Argentina, Egypt, Ethiopia, Iran, Saudi Arabia, and the United Arab Emirates.

since the Convention was adopted, raising major questions about its design and efficacy. The 2015 *Paris Agreement*, which was meant to rescue the climate regime from stalemate, has so far failed to achieve any reductions in global greenhouse emissions (UNEP 2022c). The other key convention, the 1992 *Convention on Biological Diversity* (CBD), does make the needs of ecosystems and the more-than-human prominent, but it treats Earth others as moral patients, not agential and communicative beings. The first two ten-year plans were only partially achieved, and pressures on biodiversity remain so intense that the CBD's scientific advisory body, the Intergovernmental Science-Policy Platform on Biodiversity and Ecosystem Services (IPBES), warned in 2019 that a million species (a ninth of the known species on Earth) were at risk of extinction.

The international climate regime is structured by a ruthless power politics in which the fossil fuel industry remains an influential background force, and, to the extent that it considers justice questions, they are limited to those faced by humans – primarily in 'loss and damage' and 'just transitions' out of fossil fuels and deforestation. Yet arresting dangerous climate change is a MSJ question because it is key to limiting damage to marine and terrestrial ecosystems around the world. 'Climate justice' claims that the BRICS and the global majority nations can delay reaching net zero are challenged by new research that shows that – once deforestation and land use change are included – the top five historical greenhouse polluters are the United States, China, Russia, Brazil, and Indonesia, with India and Japan in the top ten (Evans 2021; Farand 2021). A MSJ approach to climate governance should address the transitional justice needs of humans but will extend to climate-displaced people and animals and climate-vulnerable regions, such as the Sahel, East Africa, South Asia, the Caribbean, and the Pacific, who need decarbonisation to occur as rapidly as possible (Biermann & Boas 2008; Kelman 2010; Dreher & Voyer 2015). Multispecies justice will consider the myriad impacts of global heating on the Earth system, endangered keystone species like corals, and more-than-human communities across the Earth's most climatically crucial biomes, such as the polar regions and boreal and tropical forests and especially the Amazon and Indonesia (Pereira & Viola 2018, 2021). Multispecies justice-informed agreements must be grounded in the stability, health, and survival of the Earth system and all its beings and processes.

The 2015 *Paris Agreement* on climate change remains the most important international agreement to progress decarbonisation and climate change mitigation. It is also largely ineffective, at best holding the line against a total collapse in international cooperation on the climate crisis (Clémençon 2016; Allan et al. 2021). The weaknesses of the agreement are manifold but lie primarily in four areas: its consensus voting rule, which fossil fuel-dependent states have weaponised to delay action; its omission of military, maritime, and aviation emissions from mandatory national reporting and member states' inaccurate reporting of their energy, transport, and agricultural emissions and sinks; its inadequate and scientifically suspect emissions reduction pathways, which do not align with the stated goal of holding global heating to 1.5°C; and its inadequate five-yearly timetable for member states to 'rachet up' their emissions reduction commitments (Mooney et al. 2021). In total, this has created an agreement that consumes enormous government and NGO resources but has delivered little more than opacity and delay. In the meantime, the world breaks temperature and extinction records, and stable but fragile systems begin to break down.

Many governments recognise the Paris Agreement's flaws and are seeking reform. The Climate Vulnerable Forum, representing some of the world's most vulnerable African and island states, has demanded that member states be required to increase their ambition annually. This should be one crucial reform. Another more oblique tactic was the 2023 United Nations General Assembly resolution, spearheaded by Vanuatu, requesting the International Court of Justice for an advisory opinion on the responsibilities of states relating to climate change and greenhouse emissions. This resolution seeks advice on the impacts on climate-vulnerable states and people and future generations and the responsibilities of states 'to ensure the protection of the climate system and other parts of the environment'. In this way, it is weakly anthropocentric and fails to align clearly with MSJ, even as we can glimpse similarities there. The court is restricted to interpreting the question in relation to existing international law and cannot propose new law (Tigre, Bañuelos, & Bañuelos 2023). We regard the initiative as both a positive and risky one. A strong, scientifically based opinion would place more pressure on states to reduce emissions and support climate-vulnerable peoples. Yet if it is weak or flawed, it could further set back climate governance at a time when the entire Earth system and all the relations it supports are at risk.

With Paris rapidly failing, new law needs to be a more urgent priority. At Glasgow large numbers of states also signed non-binding initiatives on coal, forests, and methane in an effort to accelerate progress in key emissions sectors and precious habitats and carbon sinks (PPCA 2017; Wintour 2021; *Global Coal to Clean Power Transition Statement* 2021b; *Glasgow Leaders' Declaration on Forests and Land Use* 2021). For such voluntary initiatives to be effective, they should be translated into binding, time-bound international treaties negotiated through the General Assembly. Scientifically robust treaties to phase out and ban coal burning, deforestation, and agricultural methane are especially urgent and should be followed as soon as possible thereafter by adding oil and gas (which also contribute to methane emissions) (Abdenur 2020; Burke & Fishel 2020a; Newell & Simms 2020; Burke 2022).

The *Paris Agreement* itself can also be improved. Every document it issues now must centre the precautionary principle. States should be required to deliver increased emissions reduction commitments annually in line with the call from climate-vulnerable states and the Agreement's professed commitment to keep global heating well below 2°C. The COP must adopt rules for transparent reporting and accounting of national emissions from *all* states, along with emissions from maritime, military, and aviation. Rules and procedures for carbon markets must be ecologically rigorous and ban double-counting and projects that entrench monocultures and damage biodiversity or Indigenous rights.

The biodiversity convention has recently adopted a third ten-year plan known as the *Kunming-Montréal Global Biodiversity Framework*. It is focused on the preservation and restoration of 30 per cent of ecosystems and key habitats while centring the interests and rights of communities, especially Indigenous Peoples. Its Target 22 aims for 'the full, equitable, inclusive, effective and gender-responsive representation and participation in decision-making, and access to justice and information related to biodiversity by indigenous peoples and local communities, respecting their cultures and their rights over lands, territories, resources, and traditional knowledge' (*Convention on Biological Diversity* 2022). Yet its effectiveness will be hampered both by the principle of permanent sovereignty over natural resource, which grants states wide freedoms to overexploit nature and destroy ecosystems, and the weak accountability mechanisms in the treaty (Burke 2019; Lim 2021). The new framework implicitly acknowledges this problem. It introduces some limited enhancement of its implementation provisions – a 'global analysis' and 'global review of collective progress' enacted, we assume, in UNEP and IBPES assessments and 'voluntary peer reviews' – but they remain unenforceable soft law mechanisms that 'will be undertaken in a facilitative, non-intrusive, non-punitive manner, respecting national sovereignty, and avoiding placing undue burden on Parties' (*Convention on Biological Diversity* 2022, sec. J).

If review processes are to be inclusive of affected more-than-human communities, scientifically and ecologically rigorous analysis of progress is a good thing. Resources must be allocated to discussing weaknesses in national plans and making the peer review process as rigorous and inclusive as possible. All the Earth's major biomes must be analysed, along with all endangered and vulnerable species populations. The assessments of many species groups by the International Union for the Conservation of Nature (IUCN) are becoming dated, and knowledge of species numbers, health, movement, and distribution is patchy and often entirely missing in many states (Noss et al. 2021). The harassment and murder of environmental defenders[36] should be investigated and prosecuted and states obliged to include their progress in their reports to the Convention. Mandatory and independent reviews conducted by a United Nations agency – much as the International Atomic Energy Agency (IAEA) does regarding nuclear materials and proliferation – would be preferred, even if state assistance is welcome. Yet even this may not be enough, and a more robust architecture of criminal prosecution should be promoted. Domestically, this would take the form of independent environmental protection agencies with effective powers, ecologically robust legislation with strong criminal penalties, and independent environment courts that have created specialised knowledge and judgement. Internationally, it would take the form of an international crime of ecocide, which we explore further in Section 5.2.1.

So far we have argued for reforms that would honour human dependence on flourishing ecosystems and a stable climate. The UN leadership and treaty machinery can also be amended to ensure the inclusion of the biosphere and the more-than-human. One key initiative could be the creation of a UN High Commissioner for the Biosphere, who could play the same kind of catalytic, diplomatic, and developmental role that is discharged by existing high commissioners (or high representatives) on human rights, refugees, disarmament affairs, and the responsibility to protect. They would, consistent with current UN practice, be appointed at the Undersecretary level. This would ideally be paired with a *Universal Declaration on the Rights and Survival of the Biosphere* that would set out moral, ethical, and existential principles for new treaties and explain why the preservation and rights of the biosphere and its beings are crucial to world order and the survival of planet Earth. Language and ethics for such a declaration can already be found in multiple General Assembly resolutions and Secretary-General's reports on 'living in harmony with nature' and should reflect the core tenets of MSJ (United Nations General Assembly 2022a). The declaration could then provide the foundations of operational Earth rights covenants analogous to the *International Covenant on Civil and Political Rights* and *International Covent on Economic, Social and Cultural Rights* or concrete treaty initiatives.

Another key reform would address the very negative and planetary environmental impacts of the international customary law principle of Permanent Sovereignty Over Natural Resources (PSNR). Permanent Sovereignty Over Natural Resources should be removed from the international human rights conventions and from the *Convention on Biological Diversity* (Schuppert

---

[36] As discussed also in Section 4.

2014). It would be replaced with 'environment-neutral' language (a principle, say, of 'national economic sovereignty') that affirms the sovereign right of states to regulate, reject, phase out, and take ownership of foreign investment in their countries (subject to fair market compensation where projects are nationalised). This will preserve the original anti-imperialist aims of PSNR while removing its power to license unsustainable resource extraction with impunity. It should also include language that gives states and peoples the right to reject and close environmentally damaging projects (which may include fossil fuel extraction, mines, forestry, agriculture or plantations) where they interfere with states meeting their treaty obligations under international environmental law. We argue this to support existing efforts by many countries and the European Union to eliminate Investor State Dispute Settlement (ISDS) provisions from multilateral trade agreements, through which corporations have been able to sue democratic governments for blocking tobacco sales and fossil fuel developments in compliance with international law (Tienhaara 2018). These recommendations challenge the international property regime over Earth others and raise significant legal complexity; while they should be further researched, the regimes they seek to modify act as stubborn barriers to achieving MSJ at a planetary scale (Govind & Lim 2021; Wadiwel et al. 2023).

5.1.3 Planetary Boundaries

Planetary boundaries were first delineated by a group of scientists associated with the Stockholm Resilience Institute and are explained in three major papers (from 2009, 2015, and 2023), which develop and refine the physical model and then tighten it to set out 'safe and just' boundaries. The group's 2009 paper identified boundaries for nine key Earth system processes: 'climate change; rate of biodiversity loss (terrestrial and marine); interference with the nitrogen and phosphorus cycles; stratospheric ozone depletion; ocean acidification; global freshwater use; change in land use; chemical pollution; and atmospheric aerosol loading' (Rockström et al. 2009). Further refinements of the model included the introduction of functional biodiversity and 'biome integrity' as measures in addition to extinction rates, and environmental flows and water loss (rather than just human needs) into the freshwater use threshold (Steffen et al. 2015). New research on novel entities (pollutants such as plastics, heavy metals, chemicals and persistent organic polluting compounds) has quantified a boundary and determined that it has been exceeded (Persson et al. 2022). In the decade since the first study was done, the group has ascertained that six of the boundaries have been exceeded, many by magnitudes (Richardson et al. 2023).

Planetary boundaries define a 'safe operating space for humanity' within the Earth system that would avoid abrupt ecological change such as tipping points and Earth system 'state shifts'. In this sense they already assume some level of anthropogenic change and are weakly anthropocentric. The most recent paper, 'Safe and just Earth system boundaries', sets two boundary layers for the Earth system. A physical boundary is set to 'maintaining the resilience and stability of the Earth system', and a moral boundary set for 'minimizing exposure to significant harm to humans from Earth system change (a necessary but not sufficient condition for justice)'. We wish to note the level of moral enormity to the boundaries; one of Earth's 8.7 million species is now operating billions of simultaneous processes that collectively push beyond the limits of an entire planet, the only one known to harbour life. Urgency shadows the analysis; the authors note that global heating beyond 1.0°C above pre-industrial levels 'carries a moderate likelihood of triggering tipping elements, such as the collapse of the Greenland ice sheet or localized abrupt thawing of the boreal permafrost' (Rockström et al. 2023).

The new framework works uneasily between a primarily humanist concern with forestalling harmful 'impacts on humans, communities and countries from Earth system change' and an awareness of 'interspecies justice and Earth system stability'. The paper's authors define 'significant harm as widespread severe existential or irreversible negative impacts on countries, communities and individuals from Earth system change, such as loss of lives, livelihoods or incomes; displacement; loss of food, water or nutritional security; and chronic disease, injury or malnutrition', which becomes especially acute when affecting tens of millions of people (Rockström et al. 2023). This is no doubt profoundly important, but the rights of Earth others to flourish within the larger beings that are coral reefs, intact rainforests, or unpolluted waters remain merely a background condition.

The multispecies virtue of the planetary boundaries framework lies in its attention to the health and flourishing of the biosphere and in its estimating of hard planetary tolerances to change. The introduction of justice concerns has added ecological rigour: the authors 'propose that some boundaries be made more stringent to protect present generations and ecosystems' and that policymakers respect the more rigorous of the two. The 'safe and just' Earth system boundary for climate change has now been reduced to 1°C – which, like that for biosphere integrity, has been passed (Rockström et al. 2023). Respecting planetary boundaries means the elimination of ecologically destructive activity, major investments in restoring ecosystems globally, and rigorous limits to pollution and resource extraction that are based in sound, integrated understandings of the living balance of the Earth system. At more regional and local scales, planetary boundaries would be paired with the integrity, recovery, and flourishing of biomes and ecosystems.

There is no simple and comprehensive prescription for linking planetary boundaries to governance and policy regimes, and we do not support an authoritarian and top-down use of the framework (Pickering & Persson 2020). But it should be seen as crucial guidance that supports efforts to recentre the precautionary principle in international environmental law. The extant proposal for a global 'safe operating space treaty' is morally appealing but, in its draft form, operationally vague, lacking clarity about what kinds of institutions it would establish and what commitments from states would be required (Magalhães et al. 2016). A single treaty would face great political obstacles and could detract from more concrete efforts to create ecologically rigorous legal frameworks to address key biomes and crisis points such as pollution, fossil fuels, forests, and the oceans. An alternative argument, co-authored by many from both the planetary boundaries and Earth system governance communities, with reference to MSJ, argues for the adoption of a 'planetary commons' that address not just broad regions shared across existing national boundaries but also the biophysical systems that enable the functioning and liveability of planet Earth as a whole (Rockström et al. 2024).

Criticism of the planetary boundaries framework has come from parts of the Global South who recognise its implicit challenge to neoliberal and developmentalist models of polluting and extractivist modernisation, which the framework reveals as grossly unsustainable. In part, this issue arises because conflicts between states, corporations, and nature are crudely reframed as North–South conflicts, and it is impossible to work out 'allocations' of the safe operating space along Westphalian (national) lines (Biermann & Kim 2020, pp. 502–503). Yet when almost all the boundaries are being exceeded, what are we arguing over? We should respect the scientific and ecological validity of the boundaries and use them as crucial guidelines for operationalising limits on myriad human–ecological interactions in multiple contexts and sites, from climate change to forests, pollution, oceans, and more. They can be used to provide rigour and precaution. Societies should treat the boundaries as planetary commons shaping new forms of governance that serve the needs of more-than-humans (Galaz et al. 2012; Stephens 2022). The model's scientific accuracy and scope has been debated, modified, and integrated with social justice considerations; it is time to improve, respect, and govern the boundaries from a perspective that values MSJ across the whole Earth.

5.2 New Global Designs for Multispecies Justice

In this final part of the section, we put forward more strongly ecocentric designs that are inclusive of the more-than-human. As argued in Sections 1 and 2, we consider it a political actor deserving of inclusion because politics has sought to control, ignore, and exploit the more-than-human. We consider nature a political actor deserving of inclusion because politics has sought to control, ignore, and exploit Earth others while ignoring nature's physical power, intrinsic value, and rights to flourish (Nussbaum 2007; Burke & Fishel 2019). It is already caught up in the political. First, we argue the necessity for states to legislate a national and international crime of 'ecocide' to criminalise and prosecute the gravest abuses of ecosystems. Second, we offer a model for two new cross-national and cross-biome institutions: an Earth System Council (ESC) and the Ecoregion Assembly.

5.2.1 An International Crime of Ecocide

Broadly speaking, the crime of ecocide denotes a standalone 'great crime' akin to genocide and crimes against humanity. While ecocide is present in a few national criminal statutes and is defined in varying ways, there is no current international crime of ecocide. An important campaign, Stop Ecocide International, has been promoting the inclusion of ecocide as a fifth crime against peace in the *Rome Statute of the International Criminal Court* (Higgins, Short, & South 2012). In 2021, the campaign published a definition drafted by a high-level panel of international lawyers. It defined ecocide as 'unlawful or wanton acts committed with knowledge that there is a substantial likelihood of severe and either widespread or long-term damage to the environment being caused by those acts' (Sands et al. 2021, sec. II).

The core elements of the crime – that damage be severe and *either* widespread *or* long-term – are well-defined. They modify the definitions of these elements in the environmental war crimes provisions of international humanitarian law in a constructive and balanced way. They show a concern for ecosystem integrity, human rights to a healthy environment, and the way grave damage to ecosystems can have devastating local and planetary consequences well into the future. Scholars have expressed concern, however, that their preconditions – that the acts be 'unlawful' or 'wanton'– introduce dangerously anthropocentric biases and will make the crime impossible to prosecute (Burke & Celermajer 2021; Heller 2021; Minkova 2023). Much ecocidal damage is lawful under national law, which provides perpetrators with a protective wall of impunity. A prosecution must then clear the bar of the conduct being 'wanton', defined as 'reckless disregard for damage which would be clearly excessive in relation to the social and economic benefits anticipated' (Sands et al. 2021, sec. II). This condition assumes some ecocidal damage is acceptable in the name of human progress. According to the panel, such 'socially beneficial acts' might include building housing developments and transport links. It adds an unacceptable level of vagueness and imprecision that could open the door to permissive legal precedents and weakens the admirably ecocentric impetus of the definition with a developmentalist and anthropocentric bias that is the antithesis of MSJ.

An international law of ecocide would be valuable both for its instrumental and expressive value, but it is highly problematic to have the most important criminal law regarding the more-than-human operate within a logic where its rights are dependent on human interest. There is a simple solution to this problem: remove the 'lawful' and 'wanton' tests from the definition. Even then, the law can only deter and prosecute the worst violations of Earth rights (and only then by individuals and states (not corporations) because of the jurisdictional limits of the International Criminal Court) but it would be a valuable international capstone in the project of institutionalising MSJ (Crockett, De Sousa, & Temme 2016).

5.2.2 The Ecoregion Assembly and Earth System Council

The Ecoregion Assembly would be a new transnational governance institution with a written constitution anchored in the principles of Earth law and the Rights of Nature. These would charge the body with responsibility for the protection of the ecosystems and biodiversity in its region, and for ensuring that ecologically sensitive activities (including greenhouse emissions) that occur in its region do not do wider damage to the Earth system.

Ecoregion Assembly membership might be made up of the following: delegates of national governments with territory within the eco-region; representatives of Indigenous landowners from within the eco-region; and a substantial number of democratically elected representatives acting as proxies for more-than-human communities, who would campaign on their knowledge and track record in advocating or legislating for nature. Representatives from United Nations Environment Programme (UNEP) and the key international environmental treaty bodies should also be accredited as advisors. The number of members each assembly should have could vary by region but should be high enough to avoid risks of co-optation and corruption and to ensure responsive representation from local and regional communities, without becoming unwieldy. They should work by qualified majority voting (65 per cent) to avoid deadlock and spoiling. There should be strict controls on political donations to pre-empt corruption and capture, while providing candidates with the resources to campaign.

We propose fifteen ecoregion assemblies spread across the Earth's major biomes: the Arctic; Europe; the North Pacific; South Asia and Northeast Asia (including Myanmar, Malaysia, and the Mekong Basin); Central and

North America; the Amazon Basin; South America; South Pacific (including South-Eastern Australia); Antarctica; Archipelagic Southeast Asia (including Northern Australia and PNG); the Indian Ocean (including West Australia); Southern Africa; Central Africa; North Africa and Middle-East; and the Mediterranean and Black Seas. These ecoregions cover areas which have broadly similar climate, a limited range of ecosystem/biome types, and a manageable number of associated national states. Importantly, they would exclude states without territory in the ecoregion, eliminating a problem faced by other regional treaty bodies like the Antarctic Treaty system. They also cover key oceans and ecosystems such as the poles and the Amazon. By being geographically contiguous they can draw on regional concentrations of scientific expertise and, in most cases, successful patterns of diplomatic institutionalisation and cooperation.

We envisage that the Ecoregion Assemblies would begin by functioning more like regional organisations of states, in which resolutions and statements would provide normative guidance and force for member states or other actors, later moving to a model where Ecoregion Assemblies assume powers to legislate region-wide environmental laws and standards that must be policed and enforced. Regional member states would need to agree to yield or pool jurisdiction in environmental law-making, along the EU model, while ensuring their ability to make additional environmental laws or retain existing ones until the Ecoregion Assembly does so.

A UN ESC would play a similar role in the protection of the global ecology as the Security Council plays in the protection of international peace and security, but it would hopefully be far more representative, equitable, and effective. It would be permanently in session and, analogous to the way that Article 24 of the *UN Charter* confers on the Security Council, 'primary responsibility for the maintenance of international peace and security', the ESC would have primary responsibility for the protection and security of the global environment and Earth system. The Council would have the power to make directions to states, and its resolutions would be binding. It would also have the power to make recommendations to major treaty bodies such as the UNFCCC, the *Convention on International Trade in Endangered Species of Wild Fauna and Flora*, and the *Convention on Biodiversity*. It should not have Chapter VII-like powers to impose sanctions or authorise the use of force, although in especially severe (and we would hope rare) cases it could refer a situation to the Security Council for consideration. The Council's membership could be made up of twenty-five ambassadors: ten from United Nations member states and fifteen from the global Ecoregion Assemblies, which enables Earth others to be represented at the highest levels of the UN system. Like the Ecoregion Assembly, the Council would have a constitutional purpose to represent and protect the Earth's ecology encoded in a new establishing article of the *UN Charter*. However, absent the creation of multispecies ecoregion assemblies, we are much less certain of the value of an ESC made up solely of states given the ongoing risk of extractivist state capture.

We are aware that the success of such a system will hinge on the detailed design of the bodies, and that they represent a difficult balance between a Westphalian logic that would include states and a planetary and decolonial logic that consciously foregrounds the rights, interests, and voices of Indigenous Peoples and the more-than-human. Yet the idea is not as fantastic as it may seem: the Inuit Circumpolar Council has been in operation since 1977, and a similar experiment in transnational and multispecies governance has already been proposed in the form of the 'Andes-Amazon-Atlantic Ecological and Sociocultural Corridor' (Pereira & Terrenas 2022).

The proposed 'AAA corridor' stretches for 2.7 million km^2 from Ecuador, Peru, in the west and across Colombia, northern Brazil, and Venezuela to Guyana, Suriname, Guiana Francesca, and Brazil in the east, spanning 222 protected areas and 2,003 Indigenous territories. It has gained the support of 400 Indigenous groups, a transnational coalition of eight Amazonian NGOs, and representatives of the ministries of environment and foreign affairs of the nine Amazonian countries, but it is strongly opposed by military and extractive interests. Its aim is to 'maintain, restore and design both ecosystem and sociocultural connectivity in the northern region of the Amazon River, and promote sustainable, inclusive and interspeci[es] modes of development' (Pereira & Terrenas 2022, p. 66). According to Joana Castro Pereira and João Terrenas, the corridor is founded on an Indigenous worldview; its other name is 'the path of the Anacondas', reflecting Indigenous origin stories that humanity was born in the Atlantic mouth of the Amazon and travelled west in the form of anacondas 'along the Amazon River and its tributaries to the Andes, distributing life and giving each human group both its territory and a series of management principles, which would be essential to preserving ecological balance and the flow of vital energy in the region'. They note that by being 'conceptualised through the collaboration of different actors', the project 'recognises that any successful response to the challenges of the Anthropocene can only be achieved through reciprocal dialogue between different knowledge systems and actors. The interests of (and signals from) nature must be included within it' (Pereira & Terrenas 2022, pp. 65–69).

5.3 Conclusion

Global environmental governance has certainly done much good, but it is inadequate to the speed and complexity of the Anthropocene polycrisis. The loopholes and delay in the biodiversity and climate regimes have been especially damaging to climate-vulnerable people and Earth others. Where it has shown concern for justice, it has been for the justice of humans. We have shown that the relational demands of MSJ can be more faithfully met by a mix of reforms to existing governance regimes and the creation of new forms of transnational ecological democracy that include representatives of Indigenous Peoples and Earth others. Key principles of environmental governance must also change, and precaution must become a key approach to living within planetary boundaries. To those who might cynically emphasise the enduring presence of power politics in environmental governance, we would emphasise also that enacting MSJ at the global level is crucial for our flourishing and survival – people, states, and all Earth others.

6 Conclusion

Even the most cursory scan of the contemporary world reveals how systematic and dire is the violence and injustice committed against the more-than-human, and against groups of humans considered less-than-human. Ecological systems are collapsing under the strain of extraction, pollution, and climate change. These deteriorating ecological conditions are making survival increasingly impossible for an increasing number of species and biocultural worlds. Indigenous Peoples and peoples whose racial or class identity has been made pretext for dehumanisation are under systematic attack.

Yet the institutions purportedly established to deliver justice – through democratic decision-making, through legal regulation, through international institutions – are comprehensively failing to stem this devastating tide. In this Element, we have sought to make explicit what it is about the character of 'justice', as it is understood and practised in the dominant institutions of the contemporary world, that has rendered it impotent at best and complicit at worst.

After all, when people articulate ideas about or theories of justice, ideally, they do so to help others imagine a way of organising their worlds so that no one or no group is subjected to systematic forms of exclusion, domination, oppression, marginalisation, or violence. In turn, to the extent that they are informed by these ideals of justice, institutions are supposed to set up relationships in ways that realise this imagined world or that afford forms of reparative action when injustices inevitably occur. That the justice ideal is such an ethically compelling one makes it even more vital that ideas about and institutions of justice do not themselves inscribe arbitrary hierarchies or import assumptions that effectively create their own forms of exclusion, domination, and violence. Alas, this is precisely what dominant western theories of justice have done. They have, as we have documented in this Element, assumed that individual humans are the sole subjects of justice, that Earth others are not the type of beings, systems, or processes that can be subjects of justice – that they are indeed the 'other' that some (powerful) humans can justly use and exploit – and that justice commences when humans separate themselves from the more-than-human.

Because dominant theories of justice encode such fundamental misconceptions and unjust assumptions, no amount of tweaking or better enforcing them will be sufficient to restore broken ecological relationships or attend with care to the myriad more-than-human harms that flow from politics, law, and economics 'as usual'. What are required are understandings, institutions, and practices of governance founded on recognition of the value of the more-than-human, on acknowledgment that all lives flourish and function or flounder and fail depending on the quality of relationships in which they are embedded. Their functioning and flourishing are reliant on an appreciation that humans are a part of, not apart from, the more-than-human. At the same time, it is humans who have the ethical responsibility to re-envision and re-institutionalise justice for worlds that dominant forms of human life have devastated.

Beyond making explicit the structural failings of the political and legal institutions that currently have the greatest power to shape relations between humans and Earth others, we have sought to articulate some of the principles that might underpin transformations towards MSJ in democratic institutions and governance. More than that, we have pointed to examples, across different scales and institutional types, where these alternative principles – MSJ principles – are already at work disrupting pathological flows and relations and prefiguring more fulsome institutional overhauls. And we have planted some seeds for imagining still more radical institutional transformations where the more-than-human would, for example, be included in political decision-making, or where legal and international institutions would be structured so as to give primacy to the good and functioning of ecological relationships.

Still, in the face of the grave challenges that more-than-human worlds are facing, the rapaciousness of the capitalist and colonial forces driving their destruction and the entrenched character of the types of assumptions that normalise systematic injustice against the more-than-human, envisioning needs to be matched with strategies for actually transforming institutions. In this sense, the work we have commenced in this Element needs to be complemented by robust analyses of power, the continuing design of just governance for Earth systems, and the development of social and political movements capable of effectively advocating for and bringing about change across scales and sectors. It is partially for this reason that we approached this topic as an interdisciplinary one, both within and beyond the Earth System Governance community. All disciplines will need to work out, and work together, within the spheres that they best understand and where they have the collective capacity to bring about change.

For them – for us – to do so will require both sophisticated conceptual, institutional, and power analyses and an embodied, affective connection with the more-than-human and with all that is unfolding in our shared worlds. Dr Martin Luther King Jr. famously said that the arc of the moral universe bends towards justice; with Frederick Douglass' observation in his West India Emancipation speech of 1857, that 'power concedes nothing without a demand', we would argue that this arc does not bend, but is rather *bent* towards justice and that bending is always done by those whose experience of injustice presses upon them such that they insist that things must change. Indigenous, gender, racial, disability, and other justice transformations have always, and only, followed the demands and hard, often back-breaking work for change that members of those groups and their allies put in.

On first blush, it might seem that the same cannot be true in the case of the more-than-human. The understandings we have critiqued do not allow that they might be otherwise than moral patients, that they might be subjects that in fact speak and demand. But on closer consideration, surely the disasters unfolding are the more-than-human vociferously expressing the experience of injustice and violence, and demanding that relationships be recalibrated in ways that take account of what it will take for more-than-human worlds to function and flourish. The problem is not that there is an absence of protest or demand, but that the very institutions that inflict injustice are also structured so as to quarantine themselves from the messages and demands of the-more-than-human. At the same time, many of the humans who do have access to changing those institutions have been enculturated to be insensitive to more-than-human communications.

As matters stand, it will take more humans – and more humans across different institutional spheres – feeling themselves pressed upon by what the more-than-human is communicating to bend institutions towards justice. As we have said, Indigenous Peoples, peoples of the Global South,[37] and peoples previously colonised, enslaved, and long abused have been partners in the demands of the Earth others alongside of which they have long been mistreated and marginalised. Now, perhaps, the burdens of sensibility will be better shared, as more and more of the people for whom western liberal forms of

---

[37] By Global South, we are not making a geographic reference, but one concerning power and marginalisation. As such, certain communities within the geographic north belong to the politically and socially marginalised or the Global South, and some within the geographic south do not.

justice promised success are experiencing the impacts of treating their earthly relationships as exploitable resources in the form of food and water shortages, climate-driven disasters, and pandemics.

While a way of life for many cultures for many generations (over 65,000 years or at least 2,600 generations for Australian Aboriginal culture), multispecies concerns, relationality, and justice are relatively new to the western canon. They remain speculative and experimental in political practice, but their urgent consideration and practice is crucial for ecological functioning in a climate-challenged world.

Still, achieving a transformation of institutions along the lines we have advocated will require careful attention to the impediments that are likely to arise. Climate change may undermine more-than-human functioning faster than we expect. Complex systems are difficult to model and even harder to predict. People schooled in singular disciplines or unfamiliar with ecological approaches may reject complex approaches that address numerous and intersecting injustices. And, perhaps most direly, as conditions deteriorate, authoritarian and even fascistic responses to the climate crisis may gain the power to undermine the democratic experiments we have outlined. Even among those supportive of the immense ontological, relational, and practical shifts necessary, there are ongoing debates about basic issues of language – what to call 'nature', how to qualify the term 'justice', whether to elevate some beings over others, whether or not to include the non-living, and so on. Recognising the likelihood of these impediments ought not undermine commitment, but rather inform how bold and careful research, political and communications strategies are developed.

Here we have offered an introduction and an invitation to think further about defining and institutionalising MSJ, about both theory and practice, about relationality with the world, and particularly about human responsibility for caring for all planetary existence – for responsible custodianship and governance. We anticipate and look forward to further necessary conversations, imaginations, implementations, and more.

References

Abdenur, A. E. (2020). 'A Global Forest Treaty Is Needed Now', *PassBlue*, 10 December. www.passblue.com/2020/12/10/a-global-forest-treaty-is-needed-now/ (Accessed: 2 March 2022).

Abedisi, F. (2023). 'The Sea Casts Its Net of Justice Wide: A Speculative Judgment for What Has Been Left to the Waters of Despair', in N. Rogers and M. Maloney (eds.), *The Anthropocene Judgments Project: Futureproofing the Common Law*. Abingdon: Routledge, pp. 59–71.

Alberro, H. (2024). *Terrestrial Ecotopias: Multispecies Flourishing in and beyond the Capitalocene*. New York: Peter Lang.

Alfred, T. (1999). *Peace, Power, Righteousness*. Don Mills, ON: Oxford University Press.

Alfred, T. (2005). *Wasáse: Indigenous Pathways of Action and Freedom*. Toronto, ON: University of Toronto Press.

Alkon, A. H. and Agyeman, J. (2011). *Cultivating Food Justice: Race, Class, and Sustainability*. Cambridge, MA: MIT Press.

Alkon, A. H. and Guthman, J. (2017). *The New Food Activism: Opposition, Cooperation, and Collective Action*. Berkeley: University of California Press.

Allan, J. I., Roger, C. B., Hale, T. N., et al. (2021). 'Making the Paris Agreement: Historical Processes and the Drivers of Institutional Design', *Political Studies*, 6 October. https://doi.org/10.1177/00323217211049294.

Aristotle. (1905). *Aristotle's Politics*. Oxford: Clarendon Press.

Arvidsson, M. and Jones, E. (eds.) (2024). *International Law and Posthuman Theory*. London: Routledge.

Bader, A. (2020). 'Hostile Architecture: Our Past, Present, & Future?', *CRIT*, 86, pp. 48–51.

Balkin, A. (2015). 'Public Smog' [Performance]. Atmosphere, Earth. www.publicsmog.org.

Ball, T. (2006). 'Democracy', in A. Dobson and R. Eckersley (eds.), *Political Theory and the Ecological Challenge*. Cambridge: Cambridge University Press, pp. 131–147.

Ball, T. (2008). 'The Incoherence of Intergenerational Justice', *Inquiry: An Interdisciplinary Journal of Philosophy*, 28(1–4), pp. 321–337.

Barad, K. (2007). *Meeting the Universe Halfway: Quantum Physics and the Entanglement of Matter and Meaning*. Durham, NC: Duke University Press.

Bawaka Country, Suchet-Pearson, S., Wright, S., Lloyd, K., and Burarrwanga, L. (2013). 'Caring as Country: Towards an Ontology of Co-becoming in Natural Resource Management', *Asia Pacific Viewpoint*, 54(2), pp. 185–197. http://doi.org/10.1111/apv.12018.

Baxter, B. (2004). *A Theory of Ecological Justice*. London: Routledge.

Beck, U. (1992). *Risk Society*. London: Sage.

Biermann, F. and Boas, I. (2008). 'Protecting Climate Refugees: The Case for a Global Protocol', *Environment: Science and Policy for Sustainable Development*, 50(6), pp. 8–17. https://doi.org/10.3200/ENVT.50.6.8-17.

Biermann, F. and Kalfagianni, A. (2020). 'Planetary Justice: A Research Framework', *Earth System Governance*, 6, pp. 1–11.

Biermann, F. and Kim, R. E. (2020). 'The Boundaries of the Planetary Boundary Framework: A Critical Appraisal of Approaches to Define a "Safe Operating Space" for Humanity', *Annual Review of Environment and Resources*, 45, pp. 497–521. https://doi.org/10.1146/annurev-environ-012320-080337.

Binskin, M., Bennett, A., and Macintosh, A. (2020). Royal Commission into National Natural Disaster Arrangements – Report. Commonwealth of Australia, Canberra.

Blattner, C. E., Donaldson, S., and Wilcox, R. (2020). 'Animal Agency in Community', *Politics and Animals*, 6(1), pp. 1–22.

Boyd, D. (2017). *The Rights of Nature: A Legal Revolution that Could Save the World*. Toronto, ON: ECW Press.

Braun, B. (2005). 'Environmental Issues: Writing a More-than-Human Urban Geography', *Progress in Human Geography*, 29(5), pp. 635–650. https://doi.org.ezproxy.otago.ac.nz/10.1191/0309132505ph574pr.

Brisette, E. (2016). 'The Prefigurative Is Political: On Politics beyond "the State"', in A. C. Dinerstein (ed.), *Social Sciences for an Other Politics*. Basingstoke: Palgrave Macmillan, pp. 109–119.

Burdon, P. (2023). *The Anthropocene: New Trajectories in Law*. New York: Routledge.

Burke, A. (2019). 'Blue Screen Biosphere: The Absent Presence of Biodiversity in International Law', *International Political Sociology*, 13(3), pp. 333–351.

Burke, A. (2022). 'An Architecture for a Net Zero World: Global Climate Governance beyond the Epoch of Failure', *Global Policy*, 13(S3), pp. 24–37. https://doi.org/10.1111/1758-5899.13159.

Burke, A. and Celermajer, D. (2021). 'Human Progress Is No Excuse to Destroy Nature: A Push to Make "Ecocide" a Global Crime Must Recognise This Fundamental Truth', *The Conversation*, 31 August. http://theconversation.com/human-progress-is-no-excuse-to-destroy-nature-a-push-to-make-eco cide-a-global-crime-must-recognise-this-fundamental-truth-164594 (Accessed: 7 March 2022).

Burke, A. and Fishel, S. (2019). 'Power, World Politics, and Thing-Systems in the Anthropocene', in E. Lövbrand and F. Biermann (eds.), *Anthropocene Encounters: New Directions in Green Political Thinking*. Cambridge: Cambridge University Press, pp. 87–108. https://doi.org/10.1017/978110864 6673.005.

Burke, A. and Fishel, S. (2020a). 'A Coal Elimination Treaty 2030: Fast Tracking Climate Change Mitigation, Global Health and Security', *Earth System Governance*, 3, p. 100046. https://doi.org/10.1016/j.esg.2020.100046.

Burke, A. and Fishel, S. (2020b). 'Across Species and Borders: Political Representation, Ecological Democracy and the Non-human', in J. C. Pereira and A. Saramago (eds.), *Non-human Nature in World Politics: Theory and Practice*. Berlin: Springer Nature, pp. 33–52. https://doi.org/10.1007/978-3-030-49496-4_3.

Callicott, J. Baird. (2013). *Thinking Like a Planet: The Land Ethic and the Earth Ethic*. New York: Oxford University Press.

Caney, S. (2010). 'Climate Change, Human Rights and Moral Thresholds', in S. Humphries (ed.), *Human Rights and Climate Change*. Cambridge: Cambridge University Press, pp. 69–90.

Carter, B. and Charles, N. (2013). 'Animals, Agency and Resistance', *Journal for the Theory of Social Behavior*, 43(3), pp. 322–340.

Celermajer, D. (2021). *Summertime: Reflections on a Vanishing Future*. Sydney: Hamish Hamilton.

Celermajer, D. and McKibbin, P. (2023). 'Reimagining Relationships: Multispecies Justice as a Frame for the COVID-19 Pandemic', *Journal of Bioethical Inquiry*, 10(2), pp. 1–10.

Celermajer, D., Chatterjee, S., Cochrane, A., et al. (2020). 'Justice through a Multispecies Lens', *Contemporary Political Theory*, 19, pp. 475–512.

Celermajer, D., Schlosberg, D., Rickards, L., et al. (2021). 'Multispecies Justice: Theories, Challenges, and a Research Agenda for Environmental Politics', *Environmental Politics*, 30(1–2), pp. 119–140.

Celermajer, D., Cardoso, M., Gowers, J., et al. (2024). 'Climate Imaginaries as Praxis', *Environment and Planning E: Nature and Space*, 7(3), pp. 1015–1033.

Chakrabarty, D. (2023). *One Planet, Many Worlds: The Climate Parallax*. Waltham, MA: Brandeis University Press.

Chao, S., Bolender, K., Kirksey, E. (eds.) (2022). *The Promise of Multispecies Justice*. Durham, NC: Duke University Press.

Christoff, P. (2023). *The Fires Next Time: Understanding Australia's Black Summer*. Melbourne: Melbourne University.

Clémençon, R. (2016). 'The Two Sides of the Paris Climate Agreement: Dismal Failure or Historic Breakthrough?', *The Journal of Environment & Development*, 25(1), pp. 3–24. https://doi.org/10.1177/1070496516631362F.

Cochrane, A. (2018). *Sentientist Politics: A Theory of Global Inter-Species Justice*. Oxford: Oxford University Press.

Connolly, W. (2005). *Pluralism*. Durham, NC: Duke University Press.

Convention on Biological Diversity (2022). 'Kunming-Montreal Global Biodiversity Framework, CBD/COP/15/L.25'. United Nations.

Cooper, D. (2017). 'Transforming Markets and States through Everyday Utopias of Play', *Politica and Societa*, 2.

Cooper, D. (2019). *Feeling Like a State: Desire, Denial and the Recasting of Authority*. Durham, NC: Duke University Press.

Corkill v Forestry Commission (NSW) (1991) 73 LGRA 126.

Coulthard, G. S. (2014). *Red Skin, White Masks: Rejecting the Colonial Politics of Recognition*. Minneapolis: University of Minnesota Press.

Cover, R. M. (1985). 'The Folktales of Justice: Tales of Jurisdiction', *Capital University Law Review*, 14(2), pp. 179–203.

Coyne, B. (2017). 'The Fraught and Fishy Tale of Lungfish v The State of Queensland', in N. Rogers and M. Maloney (eds.), *Law as if Earth Really Mattered: The Wild Law Judgment Project*. New York: Routledge, pp. 56–70.

Crary, A. and Gruen, L. (2022). *Animal Crisis: A New Critical Theory*. Cambridge: Polity Press.

Cristian Rigoberto v GENEFRAN (Piatúa River case) (Judgment No. 16281-2019–00422, 5 September 2019, Multicompetent Chamber of the Provincial Court of Pastaza (Ecuador)).

Crockett, A., De Sousa, M., and Temme, J. (2016). 'International Criminal Court to Prosecute Business and Human Rights', *Insights: Herbert Smith Freehills*, 2 November. www.herbertsmithfreehills.com/insights/2016-11/international-criminal-court-to-prosecute-business-and-human-rights (Accessed: 9 November 2023).

Dahl, R. A. (1989). *Democracy and Its Critics*. New Haven, CT: Yale University Press.

Dancer, H., Holligan, B., and Howe, H. (2024). *UK Earth Law Judgments: Reimagining Law for People and Planet*. Oxford: Hart.

Dao, A. (2023). 'Swan by Her Litigation Representative Bella Donna of the Champions v Administrative Algorithmic Transformer and Minister for Immigration and Border Protection', in N. Rogers and M. Maloney (eds.),

The Anthropocene Judgments Project: Futureproofing the Common Law. Abingdon: Routledge, pp. 72–82.

Dauvergne, P. and Shipton, L. (eds.) (2023). *Global Environmental Politics in a Turbulent Era*. Cheltenham: Edward Elgar Press.

Davies, M. (2017). *Asking the Law Question*. Melbourne: Thomson Reuters.

de Fine Licht, K. (2020). '"Hostile Architecture" and Its Confederates: A Conceptual Framework for How We Should Perceive Our Cities and the Objects in Them', *Canadian Journal of Urban Research*, 29(2), pp. 1–17.

De la Cadena, M. (2015). *Earth Beings: Ecologies of Practice across Andean Worlds*. Durham, NC: Duke University Press.

De Toledo, C. (2021). Le fleuve qui voulait écrire; Les auditions du parlement de Loire, Paris, Manuella/Les Liens qui libèrent.

DeLuca, K. M. (1999). *Image Politics: The New Rhetoric of Environmental Activism*. New York: Guilford Press.

Derrida, J. (1990). 'Force of Law: The "Mystical Foundation of Authority"', *Cardozo Law Review*, 11(5–6), pp. 3–67.

de-Shalit, A. (1995). *Posterity Matters*. London: Routledge.

Dinerstein, A. C. (ed.) (2016). *Social Sciences for an Other Politics: Women Theorizing without Parachutes*. Basingstoke: Palgrave Macmillan.

Donaldson, S. (2020). 'Animal Agora: Animal Citizens and the Democratic Challenge', *Social Theory and Practice*, 46(4), pp. 709–735.

Donaldson, S. and Kymlicka, W. (2011). *Zoopolis: A Political Theory of Animal Rights*. Oxford: Oxford University Press.

Donaldson, S. and Kymlicka, W. (2015). 'Farmed Animal Sanctuaries: The Heart of the Movement?', *Politics and Animals*, 1(1), pp. 50–74.

Donaldson, S. and Kymlicka, W. (2023). 'Doing Politics with Animals', *Social Research: An International Quarterly*, 90(4), pp. 621–647.

Donoghue v Stevenson [1932] AC 562.

Donovan, J. (2006). 'Feminism and the Treatment of Animals: From Care to Dialogue', *Signs*, 31, pp. 305–329.

Dreher, T. and Voyer, M. (2015). 'Climate Refugees or Migrants? Contesting Media Frames on Climate Justice in the Pacific', *Environmental Communication*, 9(1), pp. 58–76. https://doi.org/10.1080/17524032.2014.932818.

Dryzek, J. (2002). *Deliberative Democracy and Beyond: Liberals, Critics, Contestations*. Oxford: Oxford University Press.

Dryzek, J. and Pickering, J. (2018). *The Politics of the Anthropocene*. Oxford: Oxford University Press.

Duvic-Paoli, L. (2022). 'Re-imagining the Making of Climate Law and Policy in Citizens' Assemblies', *Transnational Environmental Law*, 11(2), pp. 235–261.

Eckersley, R. (2004). *The Green State*. Cambridge, MA: MIT Press.

References

Eckersley, R. (2011). 'Representing Nature', in S. Alonso, J. Keane, and W. Merkel (eds.), *The Future of Representative Democracy*. Cambridge: Cambridge University Press, pp. 236–257.

Ehrlich, P. R. (1961). 'Has the Biological Species Concept Outlives Its Usefulness?', *Systematic Zoology*, 10(4), pp. 167–176. https://doi.org/10.2307/2411614.

Environment Council of Central Queensland v Minister for the Environment and Water (No 2) (2023) 415 ALR 318.

Environment Council of Central Queensland v Minister for the Environment and Water [2024] FCAFC 56.

Escobar, A. (2020). *Pluriversal Politics: The Real and the Possible*. Durham, NC: Duke University Press.

Estes, N. (2019). *Our History Is the Future*. London: Verso.

Esteva, G. and Prakash, M. S. (1998). *Grassroots Post-Modernism: Remaking the Soil of Cultures*. London: Zed Books.

Evans, S. (2021). 'Analysis: Which Countries Are Historically Responsible for Climate Change?', *CarbonBrief*, 5 October. www.carbonbrief.org/analysis-which-countries-are-historically-responsible-for-climate-change (Accessed: 24 February 2022).

Farand, C. (2021). 'Emerging Economies Slam Cop26 Net Zero Push as "Anti-equity"', *Climate Home News*, 20 October. www.climatechangenews.com/2021/10/20/emerging-economies-slam-cop26-net-zero-push-anti-equity/ (Accessed: 24 February 2022).

Farinea, C. (2020). 'Design for Companion Species: Developing Collaborative Multispecies Urban Environments', in J. Schröder, E. Sommariva, and S. Sposito (eds.), *Creative Food Cycles – Book 1*, pp. 245–251. https://doi.org/10.15488/10118.

Ferdinand, M. (2020). *A Decolonial Ecology: Thinking from the Caribbean World*. London: Polity.

Fishel, S. R. (2017). *The Microbial State: Global Thriving and the Body Politic*. Minneapolis: University of Minnesota Press.

Fishel, S. R. (2023). 'The Global Tree: Forests and the Possibility of a Multispecies IR', *Review of International Studies*, 49(2), pp. 223–240. https://doi.org/10.1017/S0260210522000286.

Fitz-Henry, E. (2017). 'Grief and the Inter-cultural Public Sphere', *Interface: A Journal for and about Social Movements*, 9 (2), pp. 143–161.

Fitz-Henry, E. (2024). 'Decentering the Human or Re-scaling the State?: Grassroots Movements for the 'Rights' of Nature in the United States ', in A. Nakagawa and C. Douzinas (eds.), *Non-human Rights: Critical Perspectives*. Cheltenham: Edward Elgar, pp. 205–221.

References

Fitz-Henry, E. and Klein, E. (2024). 'From Just Transitions to Reparative Transformations', *Political Geography*, 108, pp. 1–9.

Frey, R. (1983). *Rights, Killing and Suffering*. Oxford: Clarendon Press.

Future Generations v Ministry of Environment (Supreme Court of Justice Colombia, STC4360-2018, 5 April 2018).

Galaz, V., Biermann, F., Crona, B., et al. (2012). '"Planetary Boundaries": Exploring the Challenges for Global Environmental Governance', *Current Opinion in Environmental Sustainability*, 4(1), pp. 80–87. https://doi.org/10.1016/j.cosust.2012.01.006.

Garner, R. (2017). 'Animals and Democratic Theory: Beyond an Anthropocentric Account', *Contemporary Political Theory*, 16, pp. 459–477.

Gayle, D. (2023). 'Just Stop Oil Protesters Have Appeals Blocked Over Dartford Crossing Sentences', *The Guardian*, 1 August. www.theguardian.com/environment/2023/jul/31/just-stop-oil-protesters-have-appeals-blocked-over-dartford-crossing-sentences.

Gleeson, B. and Low, N. (1998). *Justice, Society, and Nature: An Exploration of Political Ecology*. New York: Routledge.

Global Alliance for the Rights of Nature (2010). 'Universal Declaration for the Rights of Mother Earth', *GARN*, 22 April. www.garn.org/universal-declaration-for-the-rights-of-mother-earth.

Goh, B. C. and Round, T. (2017). 'Wild Negligence: Donoghue v Stevenson', in N. Rogers and M. Maloney (eds.), *Law as if Earth Really Mattered: The Wild Law Judgment Project*. New York: Routledge, pp. 91–106.

Goodin, R. E. (1996). 'Enfranchising the Earth, and Its Alternatives', *Political Studies*, 44, pp. 835–849.

Goodin, R. E. (2007). 'Enfranchising All Affected Interests, and Its Alternatives', *Philosophy and Public Affairs*, 35, p. 40.

Goodin, R. E. (2016). 'Enfranchising All Subjected, Worldwide', *International Theory*, 8(3), pp. 365–389.

Gordon, B. and Roudavski, S. (2021). 'More-than-Human Infrastructure for Just Resilience: Learning from, Working with, and Designing for Bald Cypress Trees (Taxadium Distichum) in the Mississippi Delta', *Global Environment*, 14, pp. 442–474.

Gordon, R. (2015). 'Unsustainable Development', in S. Alam, S. Atapattu, C. G. Gonzalez, and J. Razzaque (eds.), *International Environmental Law and the Global South*. Cambridge: Cambridge University Press, pp. 50–73. https://doi.org/10.1017/CBO9781107295414.004.

Govind, P. and Lim, M. (2021). 'Biodiversity: The Neglected Lens for Re-imagining Property, Responsibility and Law for the Anthropocene', *7th Frontiers in Environmental Law Colloquium*. University of South Australia, 25–26

February. Adelaide: University of South Australia. www.unisa.edu.au/Calendar/7th-frontiers-in-environmental-law-colloquium/ (Accessed: 13 October 2023).

Graham, M. (1999). 'Some Thoughts about the Philosophical Underpinnings of Aboriginal Worldviews', *Worldviews: Global Religions, Culture, and Ecology*, 3(2), pp. 105–118.

Grear, A. (2020). 'Legal Imaginaries and the Anthropocene: "Of" and "For"', *Law and Critique*, 31, pp. 351–366.

Haraway, D. J. (2008). *When Species Meet*. Minneapolis: University of Minnesota Press.

Haraway, D. J. (2016). *Staying with the Trouble: Making Kin in the Chthulucene*. Durham, NC: Duke University Press.

Harper, R. A. (1923). 'The Species Concept from the Point of View of a Morphologist', *American Journal of Botany*, 10(5), pp. 229–233.

Held v Montana (Montana First Judicial District Court Lewis and Clark County, Case No CDV-2020-3-7, 14 August 2023).

Heller, K. (2021). 'Skeptical Thoughts on the Proposed Crime of "Ecocide" (That Isn't)', *Opinio Juris*, 23 June. https://opiniojuris.org/2021/06/23/skeptical-thoughts-on-the-proposed-crime-of-ecocide-that-isnt/ (Accessed: 17 October 2023).

Higgins, P., Short, D., and South, N. (2012). 'Protecting the Planet after Rio – the Need for a Crime of Ecocide', *Criminal Justice Matters*, 90(1), pp. 4–5. https://doi.org/10.1080/09627251.2012.751212.

Hooley, D. (2018). 'Political Agency, Citizenship, and Non-human Animals', *Res Publica* 24, pp. 509–530.

Hoyer, D., Bennett, J. S., Reddish, J., et al. (2023). Navigating Polycrisis: Long-Run Socio-cultural Factors Shape Response to Changing Climate. *Philosophical Transactions of the Royal Society B*, 6 November, 378(1889), 20220402.

International Rights of Nature Tribunal (2014a). *Yasuni ITT Case*. Quito, Ecuador, 17 January. www.rightsofnaturetribunal.org/cases/yasuni-itt-case.

International Rights of Nature Tribunal (2014b). *British Petroleum Deepwater Horizon Oil Spill Case*. Quito, Ecuador, 17 January. www.rightsofnaturetribunal.org/cases/british-petroleum-deepwater-horizon-oil-spill-case.

Jabr, F. (2020). 'The Social Life of Forests', *The New York Times*, 2 December. www.nytimes.com/interactive/2020/12/02/magazine/tree-communication-mycorrhiza.html.

James, W. (1977). *A Pluralistic Universe*. Cambridge, MA: Harvard University Press.

Jessup, B. and Parker, C. (2023). 'Takayna/Tarkine and the EPBC Act: From Heritage Frameworks to Habitat Thinking', in N. Rogers and M. Maloney (eds.), *The Anthropocene Judgments Project: Futureproofing the Common Law*. Abingdon: Routledge, pp. 19–38.

Johnson, H., Lewis, B., and Maguire, R. (2017). 'Whaling in the Antarctic (Australia v Japan: New Zealand Intervening)', in N. Rogers and M. Maloney (eds.), *Law as if Earth Really Mattered: The Wild Law Judgment Project*. New York: Routledge, 257–275.

Jones, P. (2014). *The Oxen at the Intersection: A Collision*. Herndon, VA: Lantern Books.

Joshi, Y. (2011). 'Respectable Queerness', *Columbia Human Rights Law Review*, 43(2), pp. 415–468.

Kauffman, C. and Martin, P. (2023). 'How Ecuador's Courts Are Giving Form and Force to Rights of Nature Norm', *Transnational Environmental Law*, 12(2), pp. 366–395.

Kelman, I. (2010). 'Hearing Local Voices from Small Island Developing States for Climate Change', *Local Environment*, 15(7), pp. 605–619. https://doi.org/10.1080/13549839.2010.498812.

Kimmerer, R. W. (2013). *Braiding Sweetgrass: Indigenous Wisdom, Scientific Knowledge and the Teachings of Plants*. Minneapolis, MN: Milkweed Editions.

Kindt, J. (2024). *The Trojan Horse and Other Stories: Ten Ancient Creatures that Make Us Human*. Cambridge: Cambridge University Press.

Kojola, E. and Pellow, D. N. (2020). 'New Directions in Environmental Justice Studies: Examining the State and Violence', *Environmental Politics*, 30(1–2), pp. 100–118.

Kolbert, E. (2014). *The Sixth Extinction: An Unnatural History*. New York: Henry Holt.

Kumar, R. (2003). 'Who Can Be Wronged?', *Philosophy & Public Affairs*, 31(2), pp. 99–118.

Kurki, M. (2020). *International Relations and Relational Universe*. Oxford: Oxford University Press.

Kvelde v NSW [2023] NSWSC 1560.

Latour, B. (2004). *The Politics of Nature*. Cambridge, MA: Harvard University Press.

Lawrence, M., Janzwood, S., and Homer-Dixon, T. (2022). 'What Is a Global Polycrisis?' *Cascade Institute Discussion Paper*; 2022–4:11. See https://cascadeinstitute.org.

Lim, M. (2020). 'Extinction: Hidden in Plain Sight – Can Stories of "The Last" Unearth Environmental Law's Unspeakable Truth?', *Griffith Law Review*, 29(4), pp. 611–642.

Lim, M. (2021). 'Repeating Mistakes: Why the Plan to Protect the World's Wildlife Falls Short', *The Conversation*, 16 July. http://theconversation.com/repeating-mistakes-why-the-plan-to-protect-the-worlds-wildlife-falls-short-164497 (Accessed: 2 March 2022).

Lim, M. (2023). 'More-than-Human Relations on the Third Rock from the Sun', in N. Rogers and M. Maloney (eds.), *The Anthropocene Judgments Project: Futureproofing the Common Law*. Abingdon: Routledge, pp. 286–300.

Lurgio, J. (2019). 'Saving the Whanganui: Can Personhood Rescue a River?', *The Guardian*, 30 November. www.theguardian.com/world/2019/nov/30/saving-the-whanganui-can-personhood-rescue-a-river.

Lyons, K. (2022). 'Rights of the Amazon in Cosmopolitical Worlds', in E. Kirksey, S. Chao, and K. Bolander (eds.), *The Promise of Multispecies Justice*. Durham, NC: Duke University Press, pp. 53–76.

Magalhães, P., Steffen, W. L., Bosselmann, Klaus, Aragao, Alexandra, and Soromenho-Marques, Viriato. (2016). *The Safe Operating Space Treaty: A New Approach to Managing Our Use of the Earth System*. Newcastle upon Tyne: Cambridge Scholars.

Magaña, P. (2022). 'The Political Representation of Nonhuman Animals', *Social Theory & Practice*, 48(4), pp. 665–690.

Malm, A. (2021). *How to Blow Up a Pipeline*. London: Verso Books.

Mancuso, S. (2021). *The Nation of Plants*. London: Profile Books.

Margulis, L. and Sagan, D. (2013). *Slanted Truths: Essays on Gaia, Symbiosis and Evolution*. Berlin: Springer Science & Business Media.

Marshall, V. (2020). 'Removing the Veil from the Rights of Nature: The Dichotomy between First Nations Customary Rights and Environmental Legal Personhood', *Australian Feminist Law Journal*, 45(2), pp. 233–248.

Martuwarra RiverOfLife, Pelizzon, A., Poelina, A., et al. (2021). 'Yoongoorrookoo: The Emergence of Ancestral Personhood', *Griffith Law Review*, 30(3), pp. 505–529.

McRuer, R. (2006). *Crip Theory: Cultural Signs of Queerness and Disability*. New York: New York University Press.

Medina, J. (2012). *The Epistemology of Resistance: Gender and Racial Oppression, Epistemic Injustice, and the Social Imagination*. Oxford: Oxford University Press.

Meijer, E. (2019). *When Animals Speak: Toward an Interspecies Democracy*. New York: New York University Press.

Micheletti, M. and Stolle, D. (2012). 'Sustainable Citizenship and the New Politics and Consumption', *The Annals of the American Academy of Political and Social Science*, 644, pp. 88–120.

Mills, C. W. (1997). *The Racial Contract*. Ithaca, NY: Cornell University Press.

Minkova, L. G. (2023). 'The Fifth International Crime: Reflections on the Definition of "Ecocide"', *Journal of Genocide Research*, 25(1), pp. 62–83. https://doi.org/10.1080/14623528.2021.1964688.

Mooney, C., Eilperin, J., Butler, D., et al. (2021). 'Countries' Climate Pledges Built on Flawed Data, Post Investigation Finds', *The Washington Post*, 7 November. www.washingtonpost.com/climate-environment/interactive/2021/greenhouse-gas-emissions-pledges-data/ (Accessed: 29 January 2023).

Morales, A. L. (2019). *Medicine Stories: Essays for Radicals*. Durham, NC: Duke University Press.

Näsström, S. (2011). 'The Challenge of the All-Affected Principle', *Political Studies*, 59(1), pp. 116–134.

Natarajan, U. and Dehm, J. (eds.) (2022). *Locating Nature: Making and Unmaking International Law*. Cambridge: Cambridge University Press.

Nedelsky, J. (2011). *Law's Relations: A Relational Theory of Self, Autonomy, and Law*. Oxford: Oxford University Press.

Newell, P. and Simms, A. (2020). 'Towards a Fossil Fuel Non-proliferation Treaty', *Climate Policy*, 20(8), pp. 1043–1054. https://doi.org/10.1080/14693062.2019.1636759.

Noss, R. F., Cartwright, J. M., Estes, D., et al. (2021). 'Improving Species Status Assessments under the U.S. Endangered Species Act and Implications for Multispecies Conservation Challenges Worldwide', *Conservation Biology*, 35(6), pp. 1715–1724. https://doi.org/10.1111/cobi.13777.

Nussbaum, M. C. (2007). *Frontiers of Justice*. Cambridge, MA: The Belknap Press of Harvard University Press.

Nussbaum, M. (2023). *Justice for Animals*. Cambridge, MA: Harvard University Press.

O'Donnell, E. and Talbot-Jones, J. (2018). 'Creating Legal Rights for Rivers: Lessons from Australia, New Zealand and India', *Ecology and Society*, 23(1), Article 7.

O'Neill, J. (1993). 'Future Generations: Present Harms', *Philosophy*, 68(35), pp. 35–51.

Opperman, R. (2022). 'The Need for a Black Feminist Climate Justice: A Case of Haunting Ecology and Eco-Deconstruction', *CR: The New Centennial Review*, 22(1), pp. 59–93.

Page, E. A. (2007). 'Intergenerational Justice of What: Welfare, Resources or Capabilities?', *Environmental Politics*, 16(3), pp. 453–469. http://doi.org/10.1080/09644010701251698.

Parfit, D. (1984). *Reasons and Persons*. Oxford: Clarendon Press.

Parry, L. (2016). 'Deliberative Democracy and Animals: Not so Strange Bedfellows', in R. Garner and S. O'Sullivan (eds.), *The Political Turn in Animal Ethics*. London: Rowman & Littlefield, pp. 137–153.

Parsons, M., Fisher, K., and Crease, R. P. (2021). *Decolonising Blue Spaces in the Anthropocene*. London: Palgrave Macmillan.

Pateman, C. (1988). *The Sexual Contract*. Cambridge: Polity Press.

Pellow, D. N. (2016). 'Toward a Critical Environmental Justice Studies', *Du Bois Review: Social Science Research on Race*, 13(2), pp. 221–236.

Pellow, D. N. (2017). *What is Critical Environmental Justice?* Cambridge: Polity Press.

Pereira, J. C. and Saramago, A. (2020). 'Introduction: Embracing Non-human Nature in World Politics', in J. C. Pereira and A. Saramago (eds.), *Non-human Nature in World Politics: Theory and Practice*. Cham: Springer International (Frontiers in International Relations), pp. 1–9. https://doi.org/10.1007/978-3-030-49496-4_1.

Pereira, J. C. and Terrenas, J. (2022). 'Towards a Transformative Governance of the Amazon', *Global Policy*, 13(3), pp. 60–75. https://doi.org/10.1111/1758-5899.13163.

Pereira, J. C. and Viola, E. (2018). 'Catastrophic Climate Change and Forest Tipping Points: Blind Spots in International Politics and Policy', *Global Policy*, 9(4), pp. 513–524. https://doi.org/10.1111/1758-5899.12578.

Pereira, J. C. and Viola, E. (2021). *Climate Change and Biodiversity Governance in the Amazon: At the Edge of Ecological Collapse?* Abingdon: Routledge.

Persson, L., Carney Almroth, B. M., Collins, C. D., et al. (2022). 'Outside the Safe Operating Space of the Planetary Boundary for Novel Entities', *Environmental Science & Technology*, 56(3), pp. 1510–1521. https://doi.org/10.1021/acs.est.1c04158.

Petrini, C. (2010). *Terra Madre: Forging a New Global Network of Sustainable Food Communities*. White River Junction, VT: Chelsea Green.

Phillips, A. (1998). *The Politics of Presence*. Oxford: Oxford University Press.

Pickering, J., Backstrand, K., and Schlosberg, D. (2020). 'Between Environmental and Ecological Democracy: Theory and Practice at the Democracy- Environment Nexus', *Journal of Environmental Policy and Planning*, 22(1), pp. 1–15.

Pickering, J. and Persson, Å. (2020). 'Democratising Planetary Boundaries: Experts, Social Values and Deliberative Risk Evaluation in Earth System Governance', *Journal of Environmental Policy & Planning*, 22(1), pp. 59–71.

Plumwood, V. (1999). 'Ecological Ethics from Rights to Recognition: Multiple Spheres of Justice for Humans, Animals and Nature', in N. Low (ed.), *Global Ethics and Environment*. New York: Routledge, pp. 188–212.

Plumwood, V. (2002). *Feminism and the Mastery of Nature*. New York: Routledge.

Poelina, A., Taylor, K. S., and Perdrisat, I. (2019). 'Martuwarra Fitzroy River Council: An Indigenous Cultural Approach to Collaborative Water Governance', *Australasian Journal of Environmental Management*, 26(3), pp. 236–254.

POLAU (2021). *Extraits de la revue de presse: La démarche du parlement du Loire*. https://drive.google.com/file/d/1ceTjKEvnyG_Q05XTiVYWUALXc T6aCjO7/view.

POLAU (2022). *Vers un parlement de Loire*. https://drive.google.com/file/d/ 1zG74Bk9YH9WzO1F_Qa1ai1V-I-2y3M24/view.

Pollastri, S., Griffiths, R., Dunn, N., et al. (2021). 'More-than-Human Future Cities: From the Design of Nature to Designing for and through Nature', *Media Architecture Biennale*, 20, pp. 23–30.

PPCA (2017). *Declaration: Powering Past Coal Alliance, Powering Past Coal Alliance (PPCA)*. www.poweringpastcoal.org/about/declaration (Accessed: 13 March 2022).

Preston, B. (2017a). 'Writing Judgments Wildly', in N. Rogers and M. Maloney (eds.), *Law as if Earth Really Mattered: The Wild Law Judgment Project*. New York: Routledge, pp. 19–27.

Preston, B. (2017b). 'Green Sea Turtles by the Representative, Meryl Streef v The State of Queensland and the Commonwealth of Australia', in N. Rogers and M. Maloney (eds.), *Law as if Earth Really Mattered: The Wild Law Judgment Project*. New York: Routledge, pp. 31–38.

Preston, B. (2018). 'The Challenges of Approaching Judging from an Earth-Centred Perspective', *Environmental and Planning Law Journal*, 35(2), pp. 218–226.

Prieto, G. (2021). 'The Los Cedros Forest Has Rights: The Ecuadorian Constitutional Court Affirms the Rights of Nature', *VerfBlog*, 10 December. https://verfassungsblog.de/the-los-cedros-forest-has-rights.

Puhakka, R., Rantala, O., Roslund, M. I., et al. (2019). 'Greening of Daycare Yards with Biodiverse Materials Affords Well-Being, Play and Environmental Relationships', *International Journal of Environmental Research and Public Health*, 16(16), p. 2948.

Rawls, J. (1971). *A Theory of Justice*. Cambridge, MA: Harvard University Press.

Re Minors Oposa v Secretary of the Department of Environment and Natural Resources (1994) 33 ILM 174.

Realpe Herrera v SENAGUA (Aquepi River case) (Judgment No. 1185-20-JP/21, 15 December 2021, Constitutional Court of Ecuador).

Red Nation (2021). *The Red Deal: Indigenous Action to Save Our Earth*. New York: Common Notions Press.

Regan, T. (2004). *The Case for Animal Rights*. Berkeley: University of California Press.
Reid, S. (2023). 'Ocean Justice: Reckoning with Material Vulnerability', *Cultural Politics*, 19(1), pp. 107–127.
Richardson, K., Steffen, W., Lucht, W., et al. (2023). 'Earth beyond Six of Nine Planetary Boundaries', *Science Advances*, 9(37), p. eadh2458. https://doi.org/10.1126/sciadv.adh2458.
Roads and Crimes Legislation Amendment Act 2022 (NSW).
Rockström, J., Steffen, W., Noone, K., et al. (2009). 'A Safe Operating Space for Humanity', *Nature*, 461(7263), pp. 472–475. https://doi.org/10.1038/461472a.
Rockström, J., Gupta, J., Qin, D., et al. (2023). 'Safe and Just Earth System Boundaries', *Nature*, 619(7968), pp. 102–111. https://doi.org/10.1038/s41586-023-06083-8.
Rockström, J., Kotzé, L., Milutinović, S., et al. (2024). 'The Planetary Commons: A New Paradigm for Safeguarding Earth Regulating Systems in the Anthropocene', *Proceedings of the National Academy of Sciences of the United States of America*, 121(5). p. e2301531121. https://doi.org/10.1073/pnas.2301531121.
Rogers, N. (2014). 'Who's Afraid of the Founding Fathers? Retelling Constitutional Law Wildly', in M. Maloney and P. Burdon (eds.), *Wild Law – In Practice*. London: Routledge, pp. 113–129.
Rogers, N. and Maloney, M. (eds.) (2017). *Law as if Earth Really Mattered: The Wild Law Judgment Project*. Abingdon: Routledge.
Rogers, N. and Maloney, M. (eds.) (2023). *The Anthropocene Judgments Project: Futureproofing the Common Law*. Abingdon: Routledge.
Roslund, M. I., Puhakka, R., Grönroos, M., et al. (2020). 'Biodiversity Intervention Enhances Immune Regulation and Health-Associated Commensal Mitobiota among Daycare Children', *Science Advances*, 6(42), p. eaba2578.
Ruru, J. (2018). 'Listening to Papatūānuku: A Call to Reform Water Law', *Journal of the Royal Society of New Zealand*, 48(2–3), pp. 215–224.
Rutz, C., Bronstein, M., Raskin, A., et al. (2023). 'Using Machine Learning to Decode Animal Communication', *Science*, 381(6654), pp. 152–155.
Sachs Olsen, C. (2019). *Socially Engaged Art and the Neoliberal City*. New York: Routledge.
Sachs Olsen, C. (2022). 'Co-Creation Beyond Humans: The Arts of Multispecies Placemaking', *Urban Planning*, 7(3), pp. 315–325.
Sand, P. H. (2008). 'The Evolution of International Environmental Law', in D. Bodansky, J. Brunnée, and E. Hey (eds.), *The Oxford Handbook of International Environmental Law, Oxford Handbooks* (2008; online edn,

Oxford Academic, 18 Sept. 2012), https://doi.org/10.1093/oxfordhb/9780199552153.013.0002.

Sands, P., Sow, F., Mackintosh, K., et al. (2021). 'Commentary and Core Text', Independent Expert Panel for the Legal Definition of Ecocide.

Sarkissian, W. (2005). 'Stories in a Park: Giving Voice to the Voiceless in Eagleby, Australia', *Planning Theory and Practice*, 6(1), pp. 103–117.

Sassen, S. (2014). *Expulsions: Brutality and Complexity in the Global Economy*. Cambridge, MA: Harvard University Press.

Sbicca, J. (2018). *Food Justice Now! Deepening the Roots of Social Struggle*. Minneapolis: University of Minnesota Press.

Schlosberg, D. (1999). *Environmental Justice and the New Pluralism*. Oxford: Oxford University Press.

Schlosberg, D. (2007). *Defining Environmental Justice: Theories, Movements, and Nature*. Oxford: Oxford University Press.

Schlosberg, D. (2013). 'Theorising Environmental Justice: The Expanding Sphere of a Discourse', *Environmental Politics*, 22(1), pp. 37–55.

Schlosberg, D. (2014). 'Ecological Justice for the Anthropocene', in M. Wissenburg and D. Schlosberg (eds.), *Animal Politics and Political Animals*. Basingstoke: Palgrave Macmillan, pp. 75–89.

Schlosberg, D. and Craven, L. (2019). *Sustainable Materialism: Environmental Movements and the Politics of Everyday Life*. Oxford: Oxford University Press.

Schmidt, J. J. (2022). 'Of Kin and System: Rights of Nature and the UN Search for Earth Jurisprudence', *Transactions – Institute of British Geographers*, 47(3), pp. 820–834. https://doi.org/10.1111/tran.12538.

Schuppert, F. (2014). 'Beyond the National Resource Privilege: Towards an International Court of the Environment', *International Theory*, 6(1), pp. 68–97. https://doi.org/10.1017/S1752971913000262.

Shah, S. (2023). 'The Animals Are Talking: What Does It Mean?', *New York Times Magazine*, 20 September. www.nytimes.com/2023/09/20/magazine/animal-communication.html.

Sheikh, H., Foth, N., and Mitchell, P. (2023). 'More-than-Human City Region Foresight: Multispecies Entanglements in Regional Governance and Planning', *Regional Studies*, 57(4), pp. 645–655.

Shen, R. (2023). *Tending Sanctuary: Multispecies Entanglements at VINE Sanctuary*. Master's Thesis. Harvard University Graduate School of Design. https://vinesanctuary.org/wp-content/uploads/tending-sanctuary.pdf.

Sierra Club v Morton, 405 US 727 (1972).

Simplican, C. S. (2015). *The Capacity Contract: Intellectual Disability and the Question of Citizenship*. Minneapolis: University of Minnesota Press.

Singer, P. (1975). *Animal Liberation*. New York: Harper Collins.

Steffen, W., Broadgate, W., Deutsch, L., et al. (2015a). 'The Trajectory of the Anthropocene: The Great Acceleration', *The Anthropocene Review*, 2(1), pp. 81–98.

Steffen, W., Richardson, K., Rockström, J., et al. (2015b). 'Planetary Boundaries: Guiding Human Development on a Changing Planet', *Science*, 347(6223), p. 736.

Stephens, T. (2022). 'Global Ocean Governance in the Anthropocene: From Extractive Imaginaries to Planetary Boundaries?', *Global Policy*, 13(S3), pp. 76–85. https://doi.org/10.1111/1758-5899.13111.

Stone, C. (1972). 'Should Trees Have Standing? – Towards Legal Rights for Natural Objects', *Southern California Law Review*, 45, pp. 450–501.

Stone, C. (2010). *Should Trees Have Standing? Law, Morality, and the Environment*. Oxford: Oxford University Press.

Sultana, F. (2021a). 'Political Ecology 1: From Margins to Center', *Progress in Human Geography*, 45(1), pp. 156–165.

Sultana, F. (2021b). 'Political Ecology II: Conjunctures, Crises, and Critical Publics', *Progress in Human Geography*, 45(6), pp. 1721–1730.

Sultana, F. (2023). 'Whose Growth in whose Planetary Boundaries? Decolonising Planetary Justice in the Anthropocene', *Geo: Geography and Environment*, 10, e00128.

Szablewska, S. and Mancini, C. (2023). 'Are Nonhuman Animals Entitled to Dignity, Privacy, and Non-exploitation? A Smart Dairy Farm of the Future', in N. Rogers and M. Maloney (eds.), *The Anthropocene Judgments Project: Futureproofing the Common Law*. Abingdon: Routledge, pp. 39–58.

Tanasescu, M. (2014). 'Rethinking Representation: The Challenge of Nonhumans', *Australian Journal of Political Science*, 49, pp. 40–53.

Tanasescu, M. (2015). 'Nature Advocacy and the Indigenous Symbol', *Environmental Values*, 24(1), pp. 105–222.

Tanasescu, M. (2023). *Understanding the Rights of Nature*. Bielefeld: Transcript.

Taylor, S. (2017). *Beasts of Burden: Animal and Disability Liberation*. New York: The New Press.

Tienhaara, K. (2018). 'Regulatory Chill in a Warming World: The Threat to Climate Policy Posed by Investor-State Dispute Settlement', *Transnational Environmental Law*, 7(2), pp. 229–250. https://doi.org/10.1017/S2047102517000309.

Tigre, M. A. and Bañuelos, J. A. C. (2023). 'The ICJ's Advisory Opinion on Climate Change: What Happens Now?', *Climate Law Blog*, 29 March. https://blogs.law.columbia.edu/climatechange/2023/03/29/the-icjs-advisory-opinion-on-climate-change-what-happens-now/ (Accessed: 9 November 2023).

Tischer, C., Kirjavainen, P., Matterne, U., et al. (2022). 'Interplay between Natural Environment, Human Microbia and Immune System: A Scoping Review of Interventions and Future Perspectives towards Allergy Prevention', *Science of the Total Environment*, 821. https://doi.org/10.1016/j.scitotenv.2022.153422.

Tschakert, P., Schlosberg, D., Celermajer, D., et al. (2021). 'Multispecies Justice: Climate-Just Futures with, for and beyond Humans', *WIREs Climate Change*, 12(2). https://doi.org/10.1002/wcc.699.

United Nations (1992a). 'United Nations Framework Convention on Climate Change', *United Nations Conference on Environment and Development*. Rio de Janeiro, 3–14 June.

United Nations (1992b). 'Rio Declaration on the Environment and Development', *United Nations Conference on Environment and Development*. Rio de Janeiro, 3–14 June.

United Nations (2007). *Declaration on the Rights of Indigenous People, RES/61/295*.

United Nations (2015). 'The Paris Agreement', *United Nations Climate Change Conference*. Paris, France, 30 November–12 December. https://unfccc.int/sites/default/files/english_paris_agreement.pdf.

United Nations (2021a). 'Glasgow Leaders' Declaration on Forests and Land Use', *United Nations Climate Change Conference*. Glasgow, United Kingdom, 31 October–13 November. Kew: The National Archives. https://ukcop26.org/glasgow-leaders-declaration-on-forests-and-land-use/ (Accessed: 2 March 2022).

United Nations (2021b). 'Global Coal to Clean Power Transition Statement', *United Nations Climate Change Conference*. Glasgow, United Kingdom, 31 October–13 November. Kew: The National Archives. https://ukcop26.org/global-coal-to-clean-power-transition-statement/ (Accessed: 14 March 2022).

United Nations (2022a). *Sustainable Development: Harmony with Nature Resolution. A/RES/77/169*. http://files.harmonywithnatureun.org/uploads/upload1295.pdf (Accessed: 5 November 2023).

United Nations (2022b). *The Human Right to a Clean, Healthy and Sustainable Environment, Res. A/76/L.75*.

United Nations (2022c). *The Closing Window: Emissions Gap Report 2022*. Nairobi: United Nations Environment Programme.

van Eeden, C., Khan, L., Osman, M. S., and Tervaert, J. W. C. (2020). 'Natural Killer Cell Dysfunction and Its Role in COVID-19', *International Journal of Molecular Sciences*, 21(17), 6351.

Van Dooren, T., Kirksey, E., and Munster, U. (2016). 'Multispecies Studies: Cultivating the Arts of Attentiveness', *Environmental Humanities*, 8(1), pp. 1–21.

VINE (2017). '2017 at VINE sanctuary', *VINE Sanctuary News*, 25 December. http://blog.bravebirds.org/archives/3177.

Wadiwel, D. (2023). 'Multispecies Economic Justice: Property in Focus', *Sydney Environmental Institute Podcast Series*. [Podcast]. www.sydney.edu.au/sydney-environment-institute/events/past-events/2022/november/multispecies-economic-justice–property-in-focus.html (Accessed: 13 October 2023).

Wang, Y., van de Wouw, M., Drogos, L., et al. (2022). 'Sleep and the Gut Microbiota in Preschool-Aged Children', *Sleep Research Society*, 45(6), pp. 1–9. https://doi.org/10.1093/sleep/zsac020.

Watene, K. (2016). 'Valuing Nature: Māori Philosophy and the Capability Approach', *Oxford Development Studies*, 44(3), pp. 287–296.

Watson, I. (1997). 'Indigenous Peoples' Law-Ways: Survival against the Colonial State', *Australian Feminist Law Journal*, 8(1), pp. 39–58.

Watson, I. (2015). *Aboriginal Peoples, Colonialism and International Law: Raw Law*. Abingdon: Routledge.

Watson, I. (2017). 'Aboriginal Laws of the Land: Surviving Fracking, Golf Courses and Drains among Other Extractive Industries', in N. Rogers and M. Maloney (eds.), *Law as if the Earth Really Mattered*. Abingdon: Routledge, 209–218.

Wenz, P. (1988). *Environmental Justice*. Albany: State University of New York Press.

Werner, J. and Lyons, S. (2020). 'The Size of Australia's Bushfire Crisis Captured in Five Big Numbers', ABC Science, 5 March. www.abc.net.au/news/science/2020-03-05/bushfire-crisis-five-big-numbers/12007716.

Whyte, K. P. (2013). 'On the Role of Traditional Ecological Knowledge as a Collaborative Concept: A Philosophical Study', *Ecological Processes*, 2(1), pp. 1–12.

Whyte, K. P. (2016). 'Our Ancestors' Dystopia Now: Indigenous Conservation and the Anthropocene', in *Routledge Companion to the Environmental Humanities*. London: Routledge, pp. 206–215.

Whyte, K. P. (2019). 'Too Late for Indigenous Climate Justice: Ecological and Relational Tipping Points', *WIREs Climate Change*, 11. https://doi.org/10.1002/wcc.603.

Whyte, K. P. and Cuomo, C. J. (2017). 'Ethics of Caring in Environmental Ethics: Indigenous and Feminist Philosophies', in S. M. Gardiner and A. Thompson (eds.), *The Oxford Handbook of Environmental Ethics*. New York: Oxford University Press, pp. 234–247.

Wilkin, J. S. (2018). *Species: The Evolution of an Idea*. Boca Raton: CRC Press. https://doi.org/10.1201/b22202.

Winne, M. (2011). *Food Rebels, Guerrilla Gardeners, and Smart-Cookin' Mamas: Fighting Back in an Age of Industrial Agriculture*. Boston, MA: Beacon Press.

Winter, C. J. (2022a). *The Subjects of Intergenerational Justice: Indigenous Philosophy, the Environment and Relationships*. London, Routledge.

Winter, C. J. (2022b). 'What's the Value of Multispecies Justice?' *Environmental Politics*, 31(2), pp. 251–257.

Winter, C. J. and Schlosberg, D. (2023). 'What Matter Matters as a Matter of Justice', *Environmental Politics*, 32(5), pp. 1–20.

Wintour, P. (2021). 'Biden Unveils Pledge to Slash Global Methane Emissions by 30%', *The Guardian*, 2 November. www.theguardian.com/environment/2021/nov/02/joe-biden-plan-cut-global-methane-emissions-30-percent (Accessed: 14 March 2022).

Wood, M. C. (2021). 'Forward', in A. R. Zelle, G. Wilson, R. Adam, H. F. Greene (eds.), *Earth Law: Emerging Ecocentric Law – A Guide for Practitioners*. Alphen aan den Rijn: Wolters Kluwer, pp. xxxiii–xl.

Woolery, L. A. (2017). 'Art-Based Perceptual Ecology: An Alternative Monitoring Method in the Assessment of Rainfall and Vegetation in a Ciénaga Community', *Artizein: Arts and Teaching Journal*, 2(6), p. 7.

World Wildlife Fund Australia (2020). *Impacts of the Unprecedented 2019–2020 Bushfires on Australian Animals*. Sydney: WWF Australia.

Wynter, S. (2003). 'Unsettling the Coloniality of Being/Power/Truth/Freedom: Towards the Human, After Man, Its Overrepresentation – An Argument', *CR: The New Centennial Review*, 3(3), pp. 257–337.

Yates, L. (2021). 'Prefigurative Politics and Social Movement Strategy: The Roles of Prefiguration in the Reproduction, Mobilisation and Coordination of Movements', *Social Movement Studies*, 69(4), pp. 1033–1052.

Yong, E. (2016). *I Contain Multitudes: The Microbes within Us and a Grander View of Life*. New York: Harper Collins.

Youatt, R. (2022). *Interspecies Politics: Nature, Borders, States*. Ann Arbor, MI: University of Michigan Press.

Youatt, R. (2023). 'Interspecies Politics and the Global Rat: Ecology, Extermination, Experiment', *Review of International Studies*, 49(2), pp. 241–257. https://doi.org/10.1017/S0260210522000201.

Young, I. (1996). 'Communication and the Other: Beyond Deliberative Democracy', in S. Benhabib (ed.), *Democracy and Difference*. Princeton, NJ: Princeton University Press, pp. 120–135.

Young, I. M. (1990). *Justice and the Politics of Difference*. Princeton, NJ: Princeton University Press.

Yunkaporta, T. and Shillingsworth, D. (2020). 'Relationally Responsive Standpoint', *Journal of Indigenous Research*, 8(2020), pp. 1–14.

About the Authors

Danielle Celermajer is Professor of Sociology at the University of Sydney, Deputy Director of the Sydney Environment Institute and leads the Multispecies Justice project. Her monographs include Sins of the Nation and the Ritual of Apology (2009) and the Prevention of Torture; An Ecological Approach (2018), both with Cambridge University Press. Her latest book *Summertime*; Reflections on a Vanishing Future (Penguin Random House, 2021) considers the more-than-human experience of climate catastrophe.

Anthony Burke is Professor of Environmental Politics and International Relations at UNSW, Canberra, Australia, and a principal at the Planet Politics Institute. His work ranges across political theory, international law, and global governance with a particular focus on climate change, biodiversity loss, Earth rights, and state ontology.

Stefanie R. Fishel is a Senior Lecturer in the School of Law and Society at the University of the Sunshine Coast. Her research takes a theoretical approach to political ecology and environmental governance with a focus on ethics, justice, and policy.

Erin Fitz-Henry is an Associate Professor of Anthropology in the School of Social and Political Sciences at the University of Melbourne. She works primarily on transnational social movements, with a particular interest in movements for the rights of nature in Ecuador, the US, and Australia. She is currently developing a new comparative research project on reparations and ecological justice in three settler-colonial contexts.

Nicole Rogers is Professor of Law at Bond University. She researches climate law, wild law, interdisciplinary climate studies and performance studies theory and the law. She co-led the Wild Law Judgment project, culminating in a collected wild law judgments in 2017. Her latest co-edited book is *The Anthropocene Judgments Project: Futureproofing the Common Law* (Routledge, 2023).

David Schlosberg is Director of the Sydney Environment Institute and Professor of Environmental Politics at the University of Sydney. His work focuses on a range of environmental justices – environmental, climate, ecological, multispecies, and just approaches to climate adaptation/resilience – as well as issues of environmental action and sustainability in everyday life.

Christine Winter (Ngati Kahungunu ki Wairoa) is a Senior Lecturer in environmental and Indigenous politics in the Politics Programme, University of Otago. She looks at ways theories of justice perpetuate injustice for some people (specifically Māori) and Earth others, focussing on critical theories of intergenerational, multispecies and planetary justice.

Cambridge Elements=

Earth System Governance

Frank Biermann
Utrecht University

Frank Biermann is Research Professor of Global Sustainability Governance with the Copernicus Institute of Sustainable Development, Utrecht University, the Netherlands. He is the founding Chair of the Earth System Governance Project, a global transdisciplinary research network launched in 2009; and Editor-in-Chief of the new peer-reviewed journal *Earth System Governance* (Elsevier). In April 2018, he won a European Research Council Advanced Grant for a research program on the steering effects of the Sustainable Development Goals.

Aarti Gupta
Wageningen University

Aarti Gupta is Professor of Global Environmental Governance at Wageningen University, The Netherlands. She is Lead Faculty and a member of the Scientific Steering Committee of the Earth System Governance (ESG) Project and a Coordinating Lead Author of its 2018 Science and Implementation Plan. She is also principal investigator of the Dutch Research Council-funded TRANSGOV project on the Transformative Potential of Transparency in Climate Governance. She holds a PhD from Yale University in environmental studies.

Michael Mason
London School of Economics and Political Science

Michael Mason is a full professor in the Department of Geography and Environment at the London School of Economics and Political Science. At LSE he is also Director of the Middle East Centre and an Associate of the Grantham Institute on Climate Change and the Environment. Alongside his academic research on environmental politics and governance, he has advised various governments and international organisations on environmental policy issues, including the European Commission, ICRC, NATO, the UK Government (FCDO), and UNDP.

About the Series

Linked with the Earth System Governance Project, this exciting new series will provide concise but authoritative studies of the governance of complex socio-ecological systems, written by world-leading scholars. Highly interdisciplinary in scope, the series will address governance processes and institutions at all levels of decision-making, from local to global, within a planetary perspective that seeks to align current institutions and governance systems with the fundamental 21st Century challenges of global environmental change and earth system transformations.

Elements in this series will present cutting edge scientific research, while also seeking to contribute innovative transformative ideas towards better governance. A key aim of the series is to present policy-relevant research that is of interest to both academics and policy-makers working on earth system governance.

More information about the Earth System Governance project can be found at: www.earthsystemgovernance.org.

Cambridge Elements

Earth System Governance

Elements in the Series

Adaptive Governance to Manage Human Mobility and Natural Resource Stress
Saleem H. Ali, Martin Clifford, Dominic Kniveton, Caroline Zickgraf and Sonja Ayeb-Karlsson

The Emergence of Geoengineering: How Knowledge Networks Form Governance Objects
Ina Möller

The Normative Foundations of International Climate Adaptation Finance
Romain Weikmans

Just Transitions: Promise and Contestation
Dimitris Stevis

A Green and Just Recovery from COVID-19?: Government Investment in the Energy Transition during the Pandemic
Kyla Tienhaara, Tom Moerenhout, Vanessa Corkal, Joachim Roth, Hannah Ascough, Jessica Herrera Betancur, Samantha Hussman, Jessica Oliver, Kabir Shahani and Tianna Tischbein

The Politics of Deep Time
Frederic Hanusch

Trade and the Environment: Drivers and Effects of Environmental Provisions in Trade Agreements
Clara Brandi and Jean-Frédéric Morin

Building Capabilities for Earth System Governance
Jochen Prantl, Ana Flávia Barros-Platiau, Cristina Yumie Aoki Inoue, Joana Castro Pereira, Thais Lemos Ribeiro and Eduardo Viola

Learning for Environmental Governance: Insights for a More Adaptive Future
Andrea K. Gerlak and Tanya Heikkila

Collaborative Ethnography of Global Environmental Governance: Concepts, Methods, Practices
Stefan C. Aykut, Max Braun and Simone Rödder

Sustaining Development in Small Islands: Climate Change, Geopolitical Security, and the Permissive Liberal Order
Matthew Louis Bishop, Rachid Bouhia, Salā George Carter, Jack Corbett, Courtney Lindsay, Michelle Scobie and Emily Wilkinson

Institutionalising Multispecies Justice
Danielle Celermajer, Anthony Burke, Stefanie R. Fishel, Erin Fitz-Henry, Nicole Rogers, David Schlosberg and Christine Winter

A full series listing is available at: www.cambridge.org/EESG